Liquid Crystal Light Modulators: Revised Edition

Edited by

Leszek R. Jaroszewicz

Faculty of Advanced Technologies and Chemistry,
Military University of Technology,
ul. gen. Sylwestra Kaliskiego 2,
00-908 Warsaw,
Poland

Liquid Crystal Light Modulators: Revised Edition

Editor: Leszek R. Jaroszewicz

ISBN (Online): 978-981-14-7022-6

ISBN (Print): 978-981-14-7020-2

ISBN (Paperback): 978-981-14-7021-9

© 2020, Bentham Books imprint.

Published by Bentham Science Publishers Pte. Ltd. Singapore. All Rights Reserved.

need for a court order if at any point you breach any terms of this License Agreement. In no event will any delay or failure by Bentham Science Publishers in enforcing your compliance with this License Agreement constitute a waiver of any of its rights.

3. You acknowledge that you have read this License Agreement, and agree to be bound by its terms and conditions. To the extent that any other terms and conditions presented on any website of Bentham Science Publishers conflict with, or are inconsistent with, the terms and conditions set out in this License Agreement, you acknowledge that the terms and conditions set out in this License Agreement shall prevail.

Bentham Science Publishers Pte. Ltd.
80 Robinson Road #02-00
Singapore 068898
Singapore
Email: subscriptions@benthamscience.net

**BENTHAM
SCIENCE**

CONTENTS

FOREWORD

The monograph concerns liquid crystal devices for modulation and switching of light. It presents rather seldom approach to the problem of liquid crystals application. Mostly, in such publications, devices for information visualization are considered, perhaps because such devices create the main segment of liquid crystals market.

The work of Professor Jaroszewicz and his coworkers concerns a more specific, sophisticated constructions, devoted to modification of various parameters of light beams. The work delivers detailed descriptions of devices dedicated to three kinds of light modification. First of them is the change of polarization state of the light beam, mostly used in pathfinders. The second one represents an electrically tuned optical filter that finds application in the detection and measurement of air pollution. The third kind of light modification device is, in principle, a quick light shutter, used first of all in welding helmets. Each of the mentioned devices modifies another parameter of the light beam: state of polarization, spectral range or transmitted light intensity. All these devices are precisely described in the review, starting with the operation principle, technical demands, design scheme and measuring methods of physical and technical parameters. A rich choice of relevant liquid crystalline materials is presented. The authors proposed many application mixtures that combined with different liquid crystal cells allowed them the construction of prototypes with unprecedented characteristics.

The presented monograph is a very useful work addressed to the scientists dealing with both basic research and liquid crystal technology. It can be useful for technicians and students as well.

Wojciech Kuczyński
Institute of Molecular Physics
Polish Academy of Sciences
ul. Smoluchowskiego 17
60-179 Poznań
Poland

PREFACE

Scientific research and custom activities in applied optics and photonics are frequently connected with the application of high-end optical elements of tailored properties. Since the liquid crystalline structures gradually become ubiquitous in optical and photonic devices, the need for a systematic description of important aspects of the contemporary material, technology, and performance of liquid crystal containing modulators is self-evident. In this book, we provide extended reports respecting our experiences in the field of the liquid crystal material studies as well as the design and technology aspects of the manufacturing of advanced liquid crystal light modulators. Our inspirations for advanced studies in mentioned subjects come from projects aiming for preparation of such adapted elements like: highly transparent, laser damage resistant liquid crystal phase modulators for space-borne laser rangefinder; high contrast, fast operating, the outdoor light shutter for eye protection, a dynamic optical filter and others. Being aware of the peculiarities of the liquid crystalline matter, moreover complexity of the design and fabrication of active optical elements, we draw up the reports providing to the reader a detailed practical input. This book is addressed to the students of technical sciences, engineers, professionals, and all the audience interested in liquid crystalline light modulators. The content of chapters partially based on our results, previously published and comprises carefully edited selection of practical information. The reader finds here the comments on chemistry and material properties of liquid crystals, custom-developed techniques of materials studies, and the rich collection of unique practical engineering solutions. The book starts with Chapter 1, providing a scope of this monograph. In Chapter 2, a new complementary method of determination of Frank elastic constants in nematic LCs is described. Chapter 3 discusses theoretical and practical aspects of the determination of key liquid crystals material parameters on examples of nematic mixtures applied for liquid crystal light modulators. Chapter 4 presents a liquid crystal polarization switch designed for the rangefinder of the space landing module. In Chapter 5, a liquid crystal spectral filter designed for visualization of air pollution is introduced. Finally, Chapter 6 presents a liquid crystal shutter for automatic welding helmet (PIAP-PS automatic). The authors would like to express sincere gratitude to all coworkers contributing to the success of scientific projects providing results and experiences presented here. Our pay-back for editing this book will be a success of the next implementations supported by its content.

Leszek Roman Jaroszewicz
Military University of Technology
ul. gen. Sylwestra Kaliskiego 2, 00-908 Warsaw
Poland

List of Contributors

Leszek R. Jaroszewicz	Faculty of Advanced Technologies and Chemistry, Military University of Technology, ul. gen. Sylwestra Kaliskiego 2, Warsaw, Poland
Paweł Perkowski	Faculty of Advanced Technologies and Chemistry, Military University of Technology, ul. gen. Sylwestra Kaliskiego 2, Warsaw, Poland
Wiktor Piecek	Faculty of Advanced Technologies and Chemistry, Military University of Technology, ul. gen. Sylwestra Kaliskiego 2, Warsaw, Poland
Zbigniew Raszewski	Faculty of Advanced Technologies and Chemistry, Military University of Technology, ul. gen. Sylwestra Kaliskiego 2, Warsaw, Poland

ACKNOWLEDGEMENTS

The first edition of this handbook (by Printing House BEL Studio Ltd., Warsaw, 2014, ISBN: 978-83-7798-141-2) was supported by The Faculty of Advanced Technologies and Chemistry of the Military University of Technology, Warsaw, Poland. This, second, corrected and completed edition has been financially supported by the Key Project POIG.01.03.01-14-016/08, carried out at the Military University of Technology in the years 2008-2015 from the Polish Ministry of Sciences and Higher Education. The editors would like to acknowledge all the people involved in this project and, more specifically, Prof. Jerzy Kędzierski and Dr. Edward Nowinowski-Kruszelnicki, for their stimulating discussions on the improvement of quality, coherence, and content presentation of chapters. Secondly, editors acknowledge the reviewer, prof. Wojciech Kuczyński (Institute of Molecular Physics, Polish Academy of Sciences, Poznań, Poland), who took part in the review process. Without the support of all the above mentioned people, this book would not have been prepared successfully.

CHAPTER 1

Introduction to Liquid Crystal Light Modulators

Zbigniew Raszewski[*]

Military University of Technology, Warsaw, Poland

Abstract: The way towards the liquid crystal modulators' design and fabrication is drawn up. The theoretical and practical aspects of material studies and implementations of tailored liquid crystals in liquid crystal modulators are discussed. The way subsequent chapters drive the reader through liquid crystal modulators' peculiarities is shown. In Chapter 2, a new complementary method of determination of Frank elastic constants in nematic LCs is described. Next, a custom method of determination of material parameters $\varepsilon_\perp$, $\Delta\varepsilon$, Δn, γ, K_{ii} of working liquid crystal nematic mixtures is discussed. Chapter 4 describes the liquid crystal polarization switch implemented in the rangefinder of the space lander module. In Chapter 5, the design and operation of a liquid crystal spectral filter for air pollution detection are discussed. Finally, Chapter 6 presents an efficient liquid crystal shutter for automatic welding helmet (PIAP-PS automatic).

Keywords: Homogeneous alignment, Homeotropic alignment, In-plane switching, Liquid crystal cell, Liquid crystal display, Liquid crystal's elastic constants, Liquid crystal filter, Liquid crystal light shutter, Nematic liquid crystals, Optical anisotropy, Permeability anisotropy, Permittivity, Refractive indices, Twisted nematic, Viscosity.

More than 120 years after F. Reinitzer's discoveries [1], due to their unique optical properties, Liquid Crystals (LCs) have found a wide range of applications, from alphanumeric displays through computer monitors to large size TV screens. The quality and reliability of displays available nowadays indicate the highest achieved level of the LC technology. Progress in LC technology has been achieved after more than 40 years of extensive studies on LC materials and electro-optical effects. The rush in LC science and technology started in 1971 – the year of M. Schadt and W. Helfrich's discovery [2] of the Twisted Nematic (TN) effect, the first electro-optical effect with the potential to be rapidly commercialized. Technology success measured in terms of a quantity of produced

---

[*] **Corresponding author Zbigniew Raszewski:** Military University of Technology, Faculty of Advanced Technologies and Chemistry, Warsaw, Poland; E-mail: zbigniew.raszewski@wat.edu.pl

Leszek R. Jaroszewicz (Ed.)

LC containing devices is strictly associated with the specific physical properties of LCs, which enable solving various technical problems in display and photonic technologies. The use of LCs in display technologies is associated with parallel, rapid development of electronic integrated circuits that enabled miniaturization of control systems and resulted in the limitation of energy consumption by consumer electronics, comprising ubiquitous Liquid Crystal Displays (LCD). Another area of importance of LC technology is the development of specialized elements of active optics, which have increasingly replaced classical ones. Such elements of adaptive optics are used for the modulation of light beam parameters like spectrum, polarization state, or propagation direction [3]. Opportunities offered by LC technology are particularly useful in the design of complex optical systems while using liquid crystal adaptive optical elements; however, there is no need to use moving mechanical parts. This allows for their easy control and reduces the implementation costs. Moreover, a flat 2D form of liquid crystal elements allows the design of systems applicable in image analysis [4, 5] and optical correlators, where simultaneous control of electric field and optical signal are applied. In recent years, photonic fibers filled with liquid crystal material, so-called Photonic Liquid Crystal Fibers (PLCFs), have been of great interest [6 - 8]. PLCs become sensitive to electric control of fiber propagation properties and the increasing transmission rate. The new area of liquid crystals' exploration is the research in the terahertz range of electromagnetic radiation [9], as well as in electrically controlled structures of metamaterials [10].

This monograph is based on the knowledge and experiences gained while selected works are done on design, fabrication, and studies of liquid crystal light modulators (LCM) [11 - 30], studies of structural and physical properties of liquid crystals [31 - 41], as well as from works concerning alignment techniques of liquid crystal layers [42 - 49]. Due to the broad area of the subject on which the authors of this monograph have been working, they focused on three following tasks done at the Military University of Technology, Warsaw, Poland (MUT):

- Liquid Crystal Cell (LCC) [11], for the space-borne rangefinder implemented at the International Space Mission "Phobos-Grunt" by Russian Federation Space Agency. The project aim was to precisely place a return module on a rugged surface of Mars' moon – Phobos. The probe with 2 LCC elements designed, manufactured, and tested at the MUT was launched on November 8th, 2011, from the Kazakhstan spaceport.
- Electrically tunable, first-order Liquid Crystal Filter (LCF) [12] destined for stations of detection and determination of air pollution in Warsaw, Poland. This filter, as well as three other liquid crystal indicators, were used in the system of monitoring and visualization of air pollution in Warsaw downtown - installed on

the arcade of the "Smyk" department store in 1996.
- Liquid Crystal Shutter (LCS) [29], for Automatic Welding Helmet (PS-automatic), was mass-produced in 1998-2005 by the Institute of Industrial Automation and Measurement in Warsaw, Poland (Safety certificate No. 244/98, 245/98 CIOP) in the amount of 100 pcs/month.

Due to the inflated technical specifications regarding aperture Φ, on- and off-switching times (τ_{on} and τ_{off}), transmittance (T) as well as wavelength range λ of modulated light for:

- LCC ($T > 95\%$, @ $\lambda = 1.062\mu m$, $\tau_{on} < 1.5$ ms, $\tau_{off} < 10.0$ ms),
- LCF ($T > 15\%$, @ $\lambda \in [0.5\mu m, 0.7\mu m]$, $\tau_{on} < 1.0$ ms, $\Phi = 160$ mm),
- LCS ($T < 0{,}0007\%$, @ $\lambda \in [0.3\mu m, 0.9\mu m]$, $\tau_{on} < 0.2$ ms);

An appropriate electro-optical effect and a modulator structure should have been chosen for each modulator separately, along with its liquid crystal working mixture of optimized properties.

Applicability of a given liquid crystal material for all-purpose applications is determined by its parameters. Among them the most important are: the dielectric anisotropy $\Delta\varepsilon$, components of permittivity tensor $\varepsilon_{\perp}$ and $\varepsilon_{\parallel}$, specifying of the type of used director field deformation and the level of the voltage U required for this purpose, optical anisotropy Δn, optimum thickness d of LC slab, rotational viscosity γ and Frank elastic constants: K_{11} for S-type deformation (splay), K_{22} for T-type deformation (twist) and K_{33} for B-type deformation (bend). Elastic constants and the dielectric anisotropy $\Delta\varepsilon$ determine the dynamic properties of the optical element indirectly.

To obtain the liquid crystal mixture with required material parameters ($\varepsilon_{\perp}$, $\Delta\varepsilon$, Δn, γ, K_{ii}) within the predefined range of working temperatures of the optical element, those parameters have to be repeatedly, and relatively quickly, determined. This is vital in a long process of composing and optimization of each working liquid crystal mixture of tailored properties and components [34]. Dielectric ($\varepsilon_{\perp}$, $\Delta\varepsilon$), optical (Δn), as well as viscosimetric parameters (γ) can be assessed by using special measuring cells under rather simple and (sometimes) automated measuring procedures. The procedures of Frank elastic constants, K_{ii}, measuring are more difficult. These constants (in particular twist constants K_{22} and bend constants K_{33}) mainly determine the switching-off times (τ_{off}) of modulators with the liquid crystal of a positive sign of the dielectric anisotropy $\Delta\varepsilon > 0$ (denoted here as PLC). The difficulty of determination of elastic constants, K_{22} and K_{33}, for PLC is associated mainly with the need to use a strong magnetic field to induce twist- and bend-type deformations in the inhomogeneously (HG) aligning layers of PLC.

It is important to know the value of the susceptibility anisotropy $\Delta\chi$ of the tested material. The use of newly developed special measuring cells with the hybrid alignment of the tested liquid crystal material [33] significantly simplifies the process of elastic constants assessment. Frank constants are now determined by using a single measuring cell of In-Plane Switching (IPS) type in which the electric field parallel to the cell walls is produced by inter-digital electrodes.

CONSENT FOR PUBLICATION

Not applicable.

CONFLICT OF INTEREST

The author(s) confirms that there is no conflict of interest.

ACKNOWLEDGEMENTS

Declared none.

REFERENCES

[1]   F. Reinitzer, "Beiträge zur Kentniss des Cholesterins", *Monatsh. Chem.,* vol. 9, pp. 421-441, 1888.
        [http://dx.doi.org/10.1007/BF01516710]

[2]   M. Schadt, and W. Helfrich, "Voltage dependent optical activity of a twisted nematic liquid crystal",
        Appl. Phys. Lett., vol. 18, no. 4, pp. 127-128, 1971.
        [http://dx.doi.org/10.1063/1.1653593]

[3]   J. Żmija, S. Kłosowicz, J. Kędzierski, E. Nowinowski-Kruszelnicki, J. Zieliński, Z. Raszewski, A.
        Walczak, and J. Parka, "Application of liquid crystals in optical processing of optical signals", *Opto-
        Electron. Rev.,* vol. 5, no. 2, pp. 93-106, 1997.

[4]   A. Walczak, E. Nowinowski-Kruszelnicki, and R. Wal, "Liquid crystal filter for polarization
        difference imaging", *Opto-Electron. Rev.,* vol. 12, no. 3, pp. 321-324, 2004.

[5]   A. Walczak, E. Nowinowski-Kruszelnicki, L. Jaroszewicz, and R. Wal, "Edge detection with liquid
        crystal polarizing filter", *Mol. Cryst. Liq. Cryst. (Phila. Pa.),* vol. 413, pp. 407-415, 2004.
        [http://dx.doi.org/10.1080/15421400490438924]

[6]   R. Dąbrowski, "New liquid crystalline materials for photonic applications", *Mol. Cryst. Liq. Cryst.
        (Phila. Pa.),* vol. 421, pp. 1-21, 2004.
        [http://dx.doi.org/10.1080/15421400490501112]

[7]   T.R. Woliński, K. Szaniawska, K. Bondarczuk, P. Lesiak, A.W. Domański, R. Dąbrowski, E.
        Nowinowski-Kruszelnicki, and J. Wójcik, "Propagation properties of photonic crystal fibers filled with
        nematic liquid crystals", *Opto-Electron. Rev.,* vol. 13, no. 2, pp. 177-182, 2005.

[8]   T.R. Woliński, A. Szymańska, T. Nasilowski, E. Nowinowski-Kruszelnicki, and R. Dąbrowski,
        "Polarization properties of liquid crystal-core optical fiber waveguides", *Mol. Cryst. Liq. Cryst. (Phila.
        Pa.),* vol. 352, pp. 361-370, 2000.
        [http://dx.doi.org/10.1080/10587250008023194]

[9]   N. Vieweg, M. Shakfa, B. Scherger, M. Mikulics, and M. Koch, "THz properties of nematic liquid
        crystals", *J. Infrared Milim. Te.,* vol. 31, no. 11, pp. 1312-1320, 2010.
        [http://dx.doi.org/10.1007/s10762-010-9721-1]

[10]   F. Zhang, L. Kang, Q. Zhao, J. Zhou, X. Zhao, and D. Lippens, "Magnetically tunable left handed metamaterials by liquid crystal orientation", *Opt. Express,* vol. 17, no. 6, pp. 4360-4366, 2009. [http://dx.doi.org/10.1364/OE.17.004360] [PMID: 19293863]

[11]   E. Nowinowski-Kruszelnicki, L. Jaroszewicz, Z. Raszewski, L. Soms, W. Piecek, P. Perkowski, J. Kędzierski, R. Dąbrowski, M. Olifierczuk, and E. Miszczyk, "Liquid crystal cell for space-borne laser rangefinder to space mission applications", *Opto-Electron. Rev.,* vol. 20, no. 4, pp. 315-322, 2012. [http://dx.doi.org/10.2478/s11772-012-0045-7]

[12]   Z. Raszewski, E. Nowinowski, J. Kędzierski, P. Perkowski, W. Piecek, R. Dąbrowski, P. Morawiak, and K. Ogrodnik, "Electrically tunable liquid crystal filters", *Mol. Cryst. Liq. Cryst. (Phila. Pa.),* vol. 525, pp. 112-127, 2010. [http://dx.doi.org/10.1080/15421401003796132]

[13]   J. Zielinski, E. Nowinowski-Kruszelnicki, R. Dąbrowski, and J. Żmija, "Twisted-nematic effect in the mixtures of p-alkylbenzoates of p-cyanophenol", *Electron. Technol.,* vol. 11, no. 3, p. 122, 1978.

[14]   E. Nowinowski-Kruszelnicki, and C. Rymarz, "Electrodynamic instabilities in liquid crystals according to the principle of conservation of moment of momentum", *J. Tech. Phys.,* vol. 19, no. 2, p. 21, 1978.

[15]   T. Cesarz, S. Kłosowicz, E. Nowinowski-Kruszelnicki, and J. Żmija, "Liquid crystal elements of laser optics. The optical isolator", *Mol. Cryst. Liq. Cryst. (Phila. Pa.),* vol. 193, pp. 19-23, 1990.

[16]   E. Nowinowski-Kruszelnicki, J. Parka, and J. Żmija, "Studies on colour effects in nematic liquid crystals", *Electron. Technol.,* vol. 12, no. 115, 1979.

[17]   A. Czapla, T.R. Woliński, E. Nowinowski-Kruszelnicki, and W.J. Bock, "A novel electrically tunable long-period fiber grating using a liquid crystal cladding layer", *Acta Phys. Pol.,* vol. 118, no. 6, pp. 1104-1107, 2010. [http://dx.doi.org/10.12693/APhysPolA.118.1104]

[18]   A. Czapla, T.R. Woliński, E. Nowinowski-Kruszelnicki, and W.J. Bock, "An electrically tunable filter based on a long-period fiber grating with a thin liquid crystal layer", *Photonics Lett. Pol.,* vol. 2, no. 3, pp. 116-118, 2010. [http://dx.doi.org/10.4302/plp.2010.3.07]

[19]   S. Ertman, T.R. Woliński, D. Pysz, R. Buczyński, E. Nowinowski-Kruszelnicki, and R. Dąbrowski, "Tunable broadband in-fiber polarizer based on photonic liquid crystal fiber", *Mol. Cryst. Liq. Cryst. (Phila. Pa.),* vol. 502, pp. 87-98, 2009. [http://dx.doi.org/10.1080/15421400902815779]

[20]   M. Sutkowski, P. Garbat, E. Nowinowski-Kruszelnicki, A. Walczak, J. Parka, and J. Woźnicki, "Polarization difference image analysis with LC filter", *Opto-Electron. Rev.,* vol. 17, no. 1, pp. 53-58, 2009. [http://dx.doi.org/10.2478/s11772-008-0035-y]

[21]   C. Tyszkiewicz, T. Pustelny, and E. Nowinowski-Kruszelnicki, "Investigation of a ferronematic cell influenced by a magnetic field", *J. Phys.,* vol. 137, pp. 161-164, 2006.

[22]   A. Walczak, and E. Nowinowski-Kruszelnicki, "Polarization sensitive liquid crystal filter for polarization difference imaging", *Proc. SPIE,* vol. 5947, pp. 76-82, 2005. [http://dx.doi.org/10.1117/12.622801]

[23]   A. Walczak, E. Nowinowski-Kruszelnicki, L. Jaroszewicz, and P. Marciniak, "Tuned liquid crystalline interferometer analysis by means of generalised Berreman matrix", *Opto-Electron. Rev.,* vol. 10, no. 1, pp. 69-73, 2002.

[24]   S. Kłosowicz, and E. Nowinowski-Kruszelnicki, "PDLC systems in elliptical capillaries", *Mol. Cryst. Liq. Cryst. (Phila. Pa.),* vol. 375, pp. 205-214, 2002. [http://dx.doi.org/10.1080/713738338]

[25] A. Walczak, E. Nowinowski-Kruszelnicki, L.R. Jaroszewicz, and P. Marciniak, "Multiresolution signal processing in liquid crystals devices", *Proc. SPIE,* vol. 4759, pp. 432-437, 2001.
[http://dx.doi.org/10.1117/12.472190]

[26] E. Nowinowski-Kruszelnicki, A. Walczak, and P. Marciniak, "The research of refractive dispersion by means of Fabry-Perot filter", *Proc. SPIE,* vol. 4759, pp. 496-504, 2012.
[http://dx.doi.org/10.1117/12.472202]

[27] S. Kłosowicz, L.R. Jaroszewicz, and E. Nowinowski-Kruszelnicki, "Optical effects in polymer-dispersed liquid crystal - fibre optic devices", *Mol. Cryst. Liq. Cryst. (Phila. Pa.),* vol. 321, pp. 323-331, 1998.
[http://dx.doi.org/10.1080/10587259808025099]

[28] E. Nowinowski-Kruszelnicki, J. Ciosek, A. Walczak, and S. Kłosowicz, "Electrically driven polarization-insensitive liquid crystal narrow-bandpass intensity modulator", *Sens. Actuators,* vol. 68, no. 1-3, pp. 316-319, 1998.
[http://dx.doi.org/10.1016/S0924-4247(98)00027-2]

[29] E. Nowinowski-Kruszelnicki, "Optical element for automatically darkening welding filters", *Proc. SPIE,* vol. 3318, pp. 519-522, 1998.
[http://dx.doi.org/10.1117/12.300038]

[30] E. Nowinowski-Kruszelnicki, E. Walczak, A. Kieżun, and L.R. Jaroszewicz, "Light transmission loss in liquid crystal waveguides", *Proc. SPIE,* vol. 3318, pp. 410-413, 1998.
[http://dx.doi.org/10.1117/12.300014]

[31] J. Kędzierski, Z. Raszewski, E. Nowinowski-Kruszelnicki, M.A. Kojdecki, W. Piecek, P. Perkowski, and E. Miszczyk, "Composite method for measurement of splay and bend nematic constants by use of single special in-plane switched cell", *Mol. Cryst. Liq. Cryst. (Phila. Pa.),* vol. 544, pp. 57-68, 2011.
[http://dx.doi.org/10.1080/15421406.2011.569273]

[32] J. Kędzierski, Z. Raszewski, M.A. Kojdecki, E. Nowinowski-Kruszelnicki, P. Perkowski, W. Piecek, E. Miszczyk, J. Zieliński, P. Morawiak, and K. Ogrodnik, "Determination of ordinary and extraordinary refractive indices of nematic liquid crystal by using wedge cells", *Opto-Electron. Rev.,* vol. 18, no. 2, pp. 214-218, 2010.
[http://dx.doi.org/10.2478/s11772-010-0009-8]

[33] E. Nowinowski-Kruszelnicki, J. Kędzierski, Z. Raszewski, L. Jaroszewicz, R. Dąbrowski, M.A. Kojdecki, W. Piecek, P. Perkowski, M. Olifierczuk, E. Miszczyk, K. Ogrodnik, P. Morawiak, and K. Garbat, "Measurement of elastic constants of nematic liquid crystal with use of hybrid in-plan--switched cell", *Opto-Electron. Rev.,* vol. 20, no. 3, pp. 255-259, 2012.
[http://dx.doi.org/10.2478/s11772-012-0027-9]

[34] E. Nowinowski-Kruszelnicki, J. Kędzierski, Z. Raszewski, L. Jaroszewicz, R. Dąbrowski, W. Piecek, P. Perkowski, M. Olifierczuk, K. Garbat, M. Sutkowski, E. Miszczyk, K. Ogrodnik, P. Morawiak, M. Laska, and R. Mazur, "High birefringence liquid crystal mixtures for lc electro-optical devices", *Opt. Appl.,* vol. 42, no. 1, pp. 167-180, 2012.

[35] E. Miszczyk, Z. Raszewski, J. Kędzierski, E. Nowinowski-Kruszelnicki, M.A. Kojdecki, P. Perkowski, W. Piecek, and M. Olifierczuk, "Interference method for determination of refractive indices of liquid crystal", *Mol. Cryst. Liq. Cryst. (Phila. Pa.),* vol. 544, pp. 22-36, 2011.
[http://dx.doi.org/10.1080/15421406.2011.569262]

[36] E. Nowinowski-Kruszelnicki, A. Walczak, and P. Marciniak, "The research of refractive dispersion by means of Fabry-Perot filter", *Proc. SPIE,* vol. 4759, pp. 496-504, 2001.
[http://dx.doi.org/10.1117/12.472202]

[37] E. Nowinowski-Kruszelnicki, A. Walczak, and P. Marciniak, "Research of chromatic dispersion by means of Fabry-Perot filter", *Opt. Appl.,* vol. 31, no. 4, pp. 751-760, 2001.

[38] A. Kieżun, L.R. Jaroszewicz, A. Walczak, and E. Nowinowski-Kruszelnicki, "Direct measurement of

refractive index profile in liquid crystal planar waveguides", *Proc. SPIE,* vol. 3745, pp. 92-98, 1999. [http://dx.doi.org/10.1117/12.357766]

[39]   A. Walczak, A. Kieżun, E. Nowinowski-Kruszelnicki, and L.R. Jaroszewicz, "New approach for a direct measurement of refractive index profile in liquid - crystalline layer", *Mol. Cryst. Liq. Cryst. (Phila. Pa.),* vol. 321, pp. 439-456, 1998. [http://dx.doi.org/10.1080/10587259808025108]

[40]   A. Walczak, E. Nowinowski-Kruszelnicki, and A. Kieżun, "Application of exact solution to anchoring energy determination", *Proc. SPIE,* vol. 3318, p. 274, 1997. [http://dx.doi.org/10.1117/12.299985]

[41]   A. Walczak, E. Nowinowski-Kruszelnicki, and A. Kieżun, "Director field in a liquid crystal: Direct measurement method", *Proc. SPIE,* vol. 3318, pp. 344-350, 1997. [http://dx.doi.org/10.1117/12.300000]

[42]   M.S. Chychłowski, E. Nowinowski-Kruszelnicki, and T.R. Woliński, "Nematic and chiral nematic liquid crystal orientation control in photonic liquid crystal fibers", *Mol. Cryst. Liq. Cryst. (Phila. Pa.),* vol. 558, pp. 28-36, 2012. [http://dx.doi.org/10.1080/15421406.2011.653676]

[43]   M.S. Chychłowski, S. Ertman, E. Nowinowski-Kruszelnicki, and T.R. Woliński, "Escaped radial and planar liquid crystal orientation inside capillaries", *Mol. Cryst. Liq. Cryst. (Phila. Pa.),* vol. 553, pp. 127-132, 2012. [http://dx.doi.org/10.1080/15421406.2011.609451]

[44]   M.S. Chychłowski, S. Ertman, E. Nowinowski-Kruszelnicki, R. Dąbrowski, and T.R. Woliński, "Comparison of different liquid crystal materials under planar and homeotropic boundary conditions in capillaries", *Acta Phys. Pol.,* vol. 120, no. 4, pp. 582-584, 2011. [http://dx.doi.org/10.12693/APhysPolA.120.582]

[45]   M.M. Chrzanowski, J. Zieliński, E. Nowinowski-Kruszelnicki, W. Piecek, M. Olifierczuk, H. Lendzion, and J. Brzeziński, "Orienting layers preparation technology for photoaligment in liquid crystal displays", *Mol. Cryst. Liq. Cryst. (Phila. Pa.),* vol. 544, pp. 170-177, 2011. [http://dx.doi.org/10.1080/15421406.2011.569399]

[46]   M.S. Chychłowski, E. Nowinowski-Kruszelnicki, and T.R. Woliński, "Liquid crystal orientation control in photonic liquid crystal fibers", *Proc. SPIE,* vol. 7753, 2011.775341 [http://dx.doi.org/10.1117/12.886044]

[47]   W.A. Stańczyk, A. Szelg, J. Kurjata, E. Nowinowski-Kruszelnicki, and A. Walczak, "Liquid crystal mono- and nano-layers covalently bonded to silicon and silica surface for alignment of LC layers", *Mol. Cryst. Liq. Cryst. (Phila. Pa.),* vol. 526, pp. 18-27, 2010. [http://dx.doi.org/10.1080/15421406.2010.485062]

[48]   S. Ertman, T.R. Woliński, and E. Nowinowski-Kruszelnicki, "Photo-induced molecular alignment in Photonic Liquid Crystal Fibers", *Proc. SPIE,* vol. 6608, p. 660809, 2007. [http://dx.doi.org/10.1117/12.739332]

[49]   S. Kłosowicz, E. Nowinowski-Kruszelnicki, and J. Żmija, "Simple method to prepare polymer dispersed liquid crystals", *Mol. Cryst. Liq. Cryst. (Phila. Pa.),* vol. 215, pp. 253-255, 1992. [http://dx.doi.org/10.1080/10587259208038532]

CHAPTER 2

Complementary Method of Assessment of Frank Elastic Constants in Nematic Liquid Crystals

Wiktor Piecek[*]

Military University of Technology, Warsaw, Poland

Abstract: A new method for quick and accurate measurements of splay (S), twist (T) and bend (B) elastic constants of nematic liquid crystals (NLC) is proposed. The main concept relies on utilizing an electric field only and on determining magnitudes of nematic elastic constants from threshold fields for Freedericksz transitions in a single, hybrid, In-Plane-Switched (IPS) cell. In Hybrid In-Plane-Switched (HIPS) cell, the deformations of an investigated LC are optically monitored while driven by three separated pairs of electrodes. Two of them are interdigitating ones. Due to the appropriate IPS electrodes and boundary conditions on them, the splay, twist and bend elastic constants can be measured without allaying the magnetic field. In this chapter, we describe the layout of HIPS measuring cells and the results of tests conducted on them using 5CB, 6CHBT (with $\Delta\varepsilon > 0$) and DE (with $\Delta\varepsilon < 0$).

Keywords: Homeotropic alignment, Homogeneous alignment, In-plane-switched cell, Nematic liquid crystals, Optical anisotropy, Permeability anisotropy, Permittivity, Refractive indices, Splay, Threshold voltages.

1. INTRODUCTION

The development of photonic technologies has resulted in intensive research of new materials, including Nematic Liquid Crystals (NLCs) of tailored properties. Material properties (generally material constants) of NLCs have to be optimized according to electro-optical effect, which is to be applied. One of the most important material parameters of NLCs are Frank constants [1]. There are different methods to measure the constants of NLCs, among which the widely used are measurements based on Freedericksz's transitions [2 - 6]. In order to obtain values of Frank constants K_{11} (for splay deformation), K_{22} (for twist

---
[*] **Corresponding author Wiktor Piecek:** Military University of Technology, Faculty of Advanced Technologies and Chemistry, Warsaw, Poland; E-mail: wiktor.piecek@wat.edu.pl

deformation) and K_{33} (for bend deformation), one has to deform a given (planar) layer of NLC with an appropriate electric E or magnetic B field (Fig. 1). In this method, the permittivity $\Delta\varepsilon$ and permeability $\Delta\chi$ anisotropies have to be previously known. An assessment of the value of permittivity anisotropy $\Delta\varepsilon$ is relatively simple and can be done with high accuracy [7 - 9]. Unfortunately, the determination of the value of permeability anisotropy $\Delta\chi$ is a very complicated matter, and it is time- and money consuming [4, 5]. The aim of our studies presented here was to elaborate and test a new, quick and relatively cheap method of specifying of a three elastic constants K_{11}, K_{22} and K_{33} by using a single hybrid measuring cell where the LC director structure is deformed by electric field only.

2. DETERMINATION OF K_{11}, K_{22} AND K_{33} IN NLCS

Quick and accurate determination of the constants K_{11}, K_{22} and K_{33} of NLCs through processes of formulation and optimization of Nematic Liquid Crystal Mixtures (LCNMs) necessary for various specialized applications is still a problem. Those measurements are usually based on the determination of critical values of electric (E_C) and magnetic (H_C) fields applied to initiate different types of Freedericksz's transitions optically monitored in planarly oriented LCNM [1 - 4, 10] (see Fig. 1).

In homogeneously (HG) or homeotropically (HT) aligned structures of tested NLCs, where the molecular director n is perfectly parallel or perpendicular to boundary surfaces, respectively, and a strong anchoring condition is met [11], a deformation is driven with splay (K_{11}), twist (K_{22}) and bend (K_{33}) constants;

$$K_{ii} = \pi^{-2} F_{ii} ,$$

(1)

may be easily calculated with recognized F_{ii} factors corresponding to experimental configurations as defined in Fig. (1). F_{ii} value is taken at the starting point of the given type (ii) of deformation. At the starting point of the n field deformation, the fields affecting NLC reach threshold (critical) values of E_C or H_C. The distribution of the director n is still homogeneous in the whole NLC slab of thickness d. Factors F can be written as:

$$F = \varepsilon_0 \Delta\varepsilon E_C^2 d^2 \quad \text{for the deformation by homogeneous electric field } E, \text{ and}$$

(2)

$$F = \mu_0 \Delta\chi H_C^2 d^2 \text{ for the deformation induced by homogeneous magnetic field } \boldsymbol{H}, \qquad \textbf{(3)}$$

where ε_0 is the permittivity and μ_0 is the susceptibility of the free space. Knowing values of $\Delta\varepsilon$ and $\Delta\chi$ anisotropies of the tested material and detecting the threshold fields E_C or H_C at given deformations realized in three different geometries (see Fig. 1), one can determine all three K_{11}, K_{22} and K_{33} elastic constants.

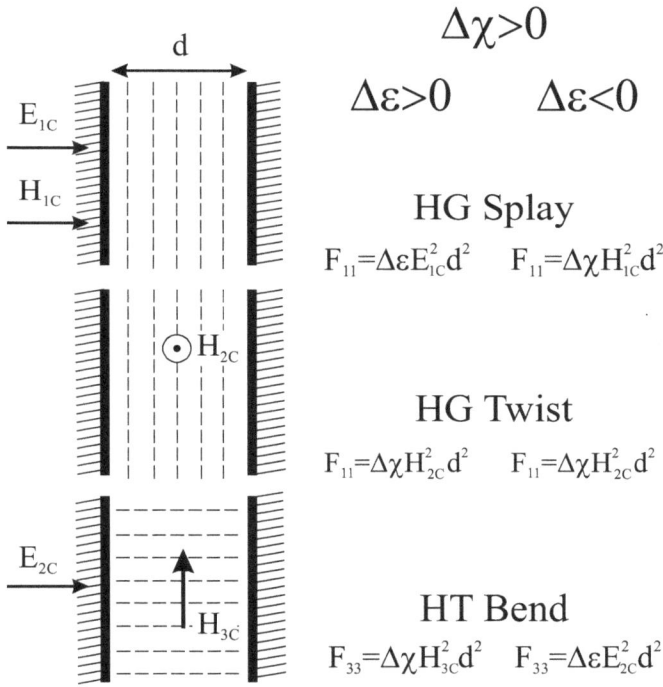

$$\Delta\chi > 0$$
$$\Delta\varepsilon > 0 \qquad \Delta\varepsilon < 0$$

HG Splay
$$F_{11} = \Delta\varepsilon E_{1C}^2 d^2 \qquad F_{11} = \Delta\chi H_{1C}^2 d^2$$

HG Twist
$$F_{11} = \Delta\chi H_{2C}^2 d^2 \qquad F_{11} = \Delta\chi H_{2C}^2 d^2$$

HT Bend
$$F_{33} = \Delta\chi H_{3C}^2 d^2 \qquad F_{33} = \Delta\varepsilon E_{2C}^2 d^2$$

Fig. (1). Geometries of NLC (with $\Delta\varepsilon > 0$ and $\Delta\varepsilon < 0$) slabs and deforming $\boldsymbol{E}$ and $\boldsymbol{H}$ fields for assessment of K_{ii} ($i = 1, 2, 3$) constants.

One can notice that for NLC of a positive dielectric anisotropy $\Delta\varepsilon > 0$ (PNLC) as well as for NLC of a negative dielectric anisotropy $\Delta\varepsilon < 0$ (NNLC), the exact value of permeability anisotropy $\Delta\chi > 0$ is required for calculations of K_{ii} elastic constants. Moreover, exact values of threshold magnetic fields H_{1C}, H_{2C}, H_{3C} are necessary for calculations of K_{11} and K_{22} for NNLC (with using of HG cells) and K_{22} and K_{33} for PNLC (with using of HT cells).

The assessment of K_{ii} in an experiment with using magnetic field $\boldsymbol{H}$ and three different measuring cells (with different alignment) with a known exact value of $\Delta\chi$ of NLC is a very complicated matter. We suggest assessment of all three constants of NLC with using a single special hybrid measuring cell and $\boldsymbol{E}$ electric field only.

If in a measuring cell filled with the NLC (for which the value of $\Delta\varepsilon$ is known), one induces homogeneous electric fields E_{3C}, E_{4C}, E_{5C}, E_{6C}, E_{7C} and E_{8C} then all three Frank constants K_{ii} can be calculated, from the Eq. (1), by putting a proper bending moment F_{ii} for given deformation (*ii*), as it is presented in Fig. (2).

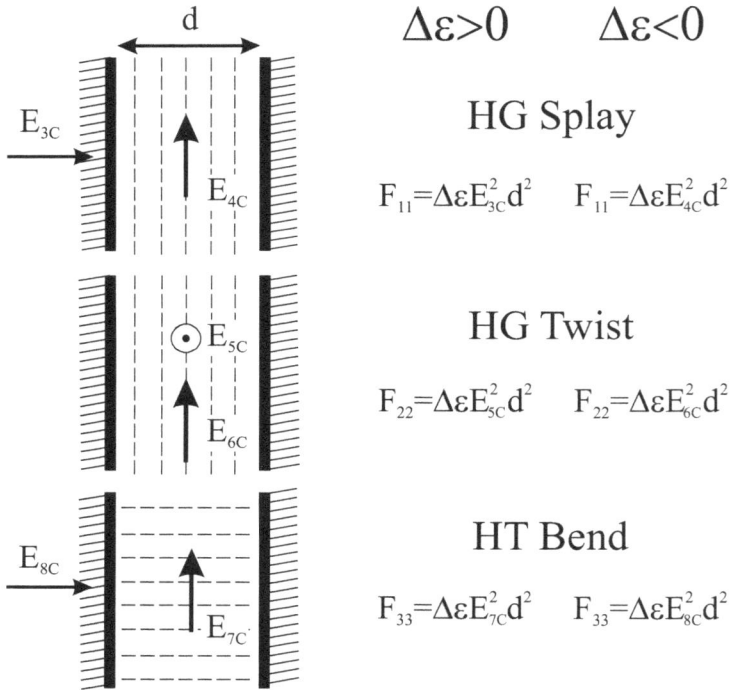

$$\Delta\varepsilon>0 \qquad \Delta\varepsilon<0$$

HG Splay

$$F_{11}=\Delta\varepsilon E_{3C}^2 d^2 \qquad F_{11}=\Delta\varepsilon E_{4C}^2 d^2$$

HG Twist

$$F_{22}=\Delta\varepsilon E_{5C}^2 d^2 \qquad F_{22}=\Delta\varepsilon E_{6C}^2 d^2$$

HT Bend

$$F_{33}=\Delta\varepsilon E_{7C}^2 d^2 \qquad F_{33}=\Delta\varepsilon E_{8C}^2 d^2$$

Fig. (2). Geometries of NLC (with $\Delta\varepsilon > 0$ and $\Delta\varepsilon < 0$) slabs for assessment of K_{ii} (*i* = 1, 2, 3) constants with $\boldsymbol{E}$ deforming field only.

In order to determine three K_{11}, K_{22} and K_{33} constants of NLC by testing its deformations induced by $\boldsymbol{E}$ field only, one has to:

• determine an accurate $\Delta\varepsilon$ value for tested NLC,

- ensure homogeneity of E_{3C}, E_{5C}, E_{7C} fields affecting PNLC or E_{4C}, E_{6C}, and E_{8C} affecting NNLC,
- determine exact values of E_{3C}, E_{5C}, and E_{7C} for PNLC or E_{4C}, E_{6C}, and E_{8C} for NNLC.

2.1. Optical Monitoring of Fréedericksz's Transitions Induced by Electric Fields E in NLCs

To replace in the Eq. (1), the bending moment F induced by magnetic field H by the bending moment from Eq. (3) induced by homogenous electric field E, the homogeneity of deforming electric fields E_{4C}, E_{5C}, E_{6C}, and E_{7C} (see Fig. 2) in the whole NLC slab of thickness d is required. Since E_{4C}, E_{5C}, E_{6C}, and E_{7C} have to lie in a plane parallel to the boundary surfaces, such fields can be induced by applying a voltage to inter-digital electrodes of In-Plane-Switching type (IPS) cell. The schematic diagram of the IPS cell is presented in Fig. (3).

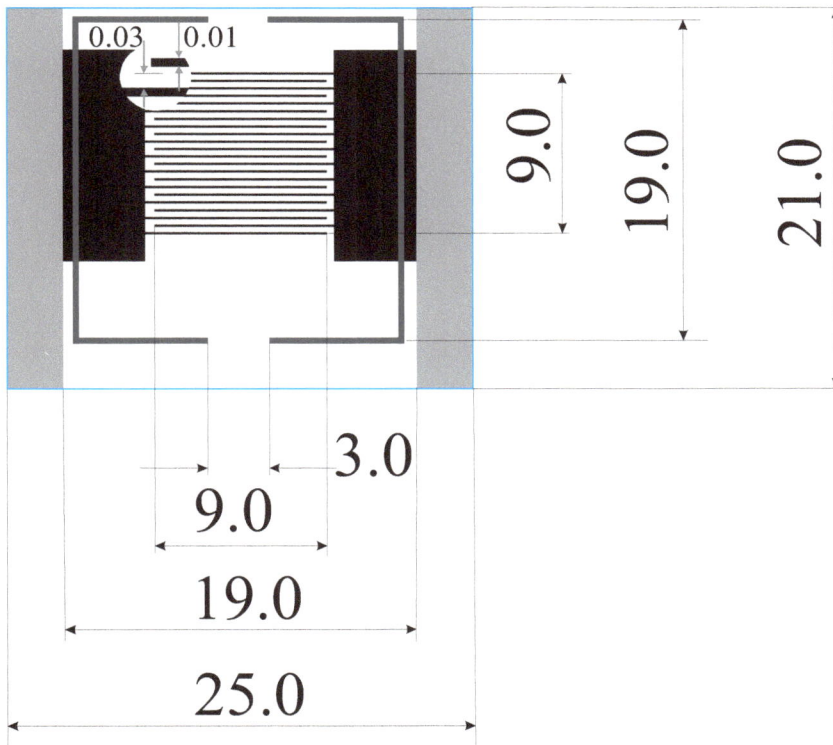

Fig. (3). Layout of a measuring IPS cell. 300 pairs of electrodes of width $b = 10$ μm separated from each other by a gap $l = 20$ μm.

In order to determine specifications of a measuring IPS cell (of a cell gap d) with interdigitating electrodes of width b and electrode gap l which meet the condition of homogeneity of deforming electric field, a computer simulation of electric field potential surrounding IPS planar electrodes was carried out. The simulation was done with an assumption that electrodes are planar, infinitely long and are of negligible thickness. As a result of these calculations, authors came up with the following conclusions;

- in order to obtain wide areas with a homogeneous electric field, stripes of electrodes should be narrow in comparison with the gap between them ($b = \frac{1}{4} l$),
- in order to obtain a moderately homogeneous distribution of E field in the cell with cell gap d, one should design and manufacture measuring cells which will meet the condition $d/(b+l) \leq 0.1$,
- the width b of the electrodes should be small enough, and the length (L) should be big enough, to prevent their edges from interfering with electric field E inside the cell.

2.1.1. Experimental Determination of Homogeneity Conditions of E Fields in IPS Cells

In order to determine the specification of parameters for measuring IPS cells with acceptable E field homogeneity, the following was done:

a. Electrodes topography was developed; masks for photolithography with the various gradation of period (c) of interdigitating, IPS electrodes were manufactured to build different types of measuring cells; chrome masks were manufactured with parameters as follows:
   - electrode width $b = 10$ μm, gap $l = 20$ μm (period $c = 30$ μm),
   - electrode width $b = 10$ μm, gap $l = 40$ μm (period $c = 50$ μm),
   - electrode width $b = 20$ μm, gap $l = 30$ μm (period $c = 50$ μm),
   - electrode width $b = 30$ μm, gap $l = 50$ μm (period $c = 80$ μm),
   - electrode width $b = 50$ μm, gap $l = 70$ μm (period $c = 120$ μm),
   - electrode width $b = 100$ μm, gap $l = 100$ μm (period $c = 200$ μm),
b. The following types of measuring cell were manufactured for determining K_{11}, K_{22}, K_{33} Frank elastic constants of NLC with using E field only:
   - Three series - HGP, HGR, HT of measuring cells. In series HGP and HGR, homogeneous alignment was induced by means of polymer (Sunever SE-130 by NISSAN Chem. Corp.) rubbed with a textile YA-19R (by Yoshikawa Chem. Comp.). In HGP cells, the rubbing direction was perpendicular, and in HGR, the direction was parallel to electrodes direction and anti-parallel on

the opposite, passive surfaces. In series HT, the homeotropic alignment was produced by means of polymer SE-1211 by Sunever.

○ Three series - KHGP, KHGR, KHT of wedge-shaped cells were built with the cell gap gradient. In those cells, a specific cell gap d at a threshold line of the NLC deformation was calculated on the basis of the wedge angle α.

○ Three series - CHGP, CHGR, CHT of cylindrical cells were built with a cell gap gradient. In those cells, cell gap d at a threshold line of the deformed NLC was calculated on the basis of a curvature radius ρ of the cylindrical lens which was the top surface of the measuring cell.

The measuring cells of series described above have a single, planar IPS electrodes set with characteristic periods: $b = 10$ μm, $l = 20$ μm; $b = 10$ μm, $l = 40$ μm; $b = 20$ μm, $l = 30$ μm; $b = 30$ μm, $l = 50$ μm; $b = 50$ μm, $l = 70$ μm; $b = 100$ μm, $l = 100$ μm. We prepared cells with three different cell gaps: $d = 5$ μm, $d = 10$ μm and $d = 15$ μm. The series of cells were manufactured by using a full professional LCD technology in the laboratory of the Military University of Technology (MUT), Warsaw, Poland.

c. Cells with planar alignment layers HGP, HGR and HG (of various cell gaps d and different periods of inter-digital electrodes), wedge-shaped cells KHGP, KHGR and KHT (with different wedge angles α) and cylindrical cells CHGP, CHGR and CHT (with different curvature radius ρ) were filled with NLCs well-known from the literature [5, 6, 10] of acronyms 6CHBT and 5CB with $\Delta\varepsilon > 0$ and with MBBA and DE with $\Delta\varepsilon < 0$. Frank elastic constants K_{ii} were measured for them.

From the analysis of the preliminary results, it was noted that in thin, planar cells of IPS, the homogeneity condition of the electric E field was the best in the inter-electrodes area. Therefore, further studies were carried out exclusively on planar cells.

2.1.2. Determination of Twist Elastic Constant K_{22}

In the laboratory coordinate system $Oxyz$ (see Fig. **4**), let us consider a beam of unpolarized light of intensity I_0, propagating along the Oz axis, incident normally on a linear polarizer P. Next, the linearly polarized beam travels through a non-absorbing, birefringent medium (NLC slab) and reaches an analyzer A. Let us consider an NLC slab of thickness d of a homogeneous alignment placed between two boundary surfaces S_1 and S_2 as a birefringent object. Both the boundary surfaces are rubbed so that the condition of a strong anchoring with the director parallel to the boundary surface is met (Tilt Bias Angle (TBA) equals 0). The rubbing direction is defined by an angle γ in the Oxy plane (see Fig. **4**). On one cell's surface S_2 the interdigitating electrodes of width b, separated by gap l, were etched in a transparent, conductive Indium Tin Oxide ITO layer. The direction of interdigitating electrodes coincides with the direction of the Ox axis of the system of coordinates. NLC layer formed in this way is described by a homogeneous field of the director $\boldsymbol{n}$ which is parallel to boundary surfaces S_1 and S_2 and forms an angle δ with the Ox axis. Polarizer's and analyzer's transmittance equals τ. Their planes, as well as planes of the electrodes, are parallel to the Oxy plane.

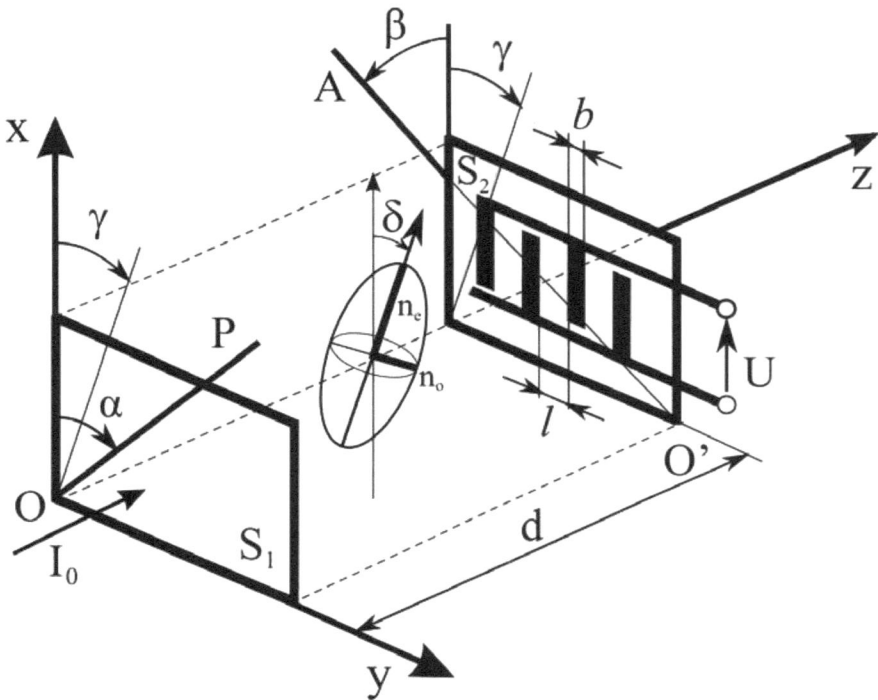

Fig. (4). NLC slab of a thickness d placed between two boundary surfaces S_1 and S_2 Rubbing direction is defined with an angle γ. On one cell's surface (S_2), transparent conducting IPS electrodes were etched in the ITO layer. The orientation of the NLC layer is described by the director $\boldsymbol{n}$ which forms an angle δ with the Ox axis. OP and OA directions, describing axes of polarizer and analyzer, form angles α and β with the Ox axis respectively.

When the director $\boldsymbol{n}$ of an NLC layer, with a positive optical anisotropy Δn, forms an angle with the Ox axis, the optical indicatrix of this layer is described by an ellipsoid. Its long axis is the optical axis of the NLC slab and it is collinear with the direction of the NLC's molecular director $\boldsymbol{n}$.

According to the above, assuming that rubbing direction (hence the director **n**) is parallel to the *Ox* axis, the light intensity *I* observed after passing through the polarizer P, NLC birefringent layer and finally the analyzer A, can be calculated as follows [12 - 14]:

$$I = 2I_o \tau^2 \left(cos^2(\alpha - \beta) - sin\,2(\alpha - \delta)\,sin\,2(\beta + \delta)\,sin^2 \frac{\pi d \Delta n}{\lambda} \right), \qquad (4)$$

where λ is the wavelength of the light and the transmittance of $\tau = \frac{1}{2}$ of an ideal polarizer is assumed.

2.1.2.1. Optical Monitoring of Twist Deformation Induced by Electric Field E for PNLC ($\Delta \varepsilon > 0$)

Let's place PNLC with $\Delta \varepsilon > 0$ and $\Delta n > 0$ in IPS cell in which a rubbing direction coincides with the *Ox* axis ($\gamma = 0$) and let's assume that the voltage *U* applied to electrodes equals to zero. In this case, the long axis of optical indicatrix (n_e) is collinear with that of the ellipsoid of permittivity ($\varepsilon_{||}$) and with the director **n**. When $\gamma = 0$ and polarizers are crossed ($\alpha = 0$ and $\beta = -\pi/2$), the light intensity *I* passing through the system, (see Eq. (4)), equals zero. An increasing voltage *U* induces electric field **E** in a PNLC layer bound by surfaces S_1 and S_2. The condition of a strong anchoring is assumed ($W = \infty$) with TBA equal 0.

In some cases, when $l \gg b$ and $l \gg d$, the electric field **E** produced by IPS electrodes may be considered homogeneous. Vector **E**, of a magnitude $E = U/l$, is parallel to S_1 and S_2 surfaces and perpendicular to electrodes. Increasing voltage *U* reaches the threshold value U_C, hence an electric field $E_C = U_C/l$ is induced, at which the twist deformation starts. However, it is worth mentioning that during the twist deformation, when voltage *U* further increases, at $U > U_C$, the mean optical indicatrix of the deformed PLC layer changes not only its shape but also spatial orientation described by the angle $\delta(U)$. At the very beginning of the deformation, the mentioned indicatrix is still an ellipsoid of revolution with major and minor semi-axes n_e and n_o almost the same as the values of extraordinary n_e and ordinary n_o refractive indices of the undeformed PNLC layer, but its long axis forms the noticeable angle $\delta(U) > 0$.

Thus, for twist deformation of PNLC layer (with $\Delta \varepsilon > 0$) arranged by rubbing with $\gamma = 0$ and placed between crossed polarizers ($\alpha = 0$ and $\beta = -\pi/2$), equation (4) can be written as [15]:

$$I = 2I_o \tau^2 \sin^2 2\delta(U)[1 - \sin^2 \frac{\pi d \Delta n(U)}{\lambda}] \qquad (5)$$

Twist deformation can be monitored by light when $\gamma = 0$, and polarizers form angles $\alpha = \pi/4$ and $\beta = -\pi/4$. In this case, equation (4) takes the form of:

$$I = 2I_o \tau^2 \cos^2 2\delta(U) \sin^2 \frac{\pi d \Delta n(U)}{\lambda} \qquad (6)$$

From equations (5) and (6), it can be seen that the best conditions for light monitoring of the twist deformation in IPS cell of PNLC layer with $\Delta\varepsilon > 0$ oriented by rubbing with $\gamma = 0$ are met when:

• $\alpha = 0$ and $\beta = -\pi/2$, and light is monochromatic with wavelength $\lambda = (d\Delta n)/k$, ($k = 1, 2, 3..$)

or

• $\alpha = \pi/4$ and $\beta = -\pi/4$, and light is monochromatic with a wavelength $\lambda = (d\Delta n)/(2k+1)$, ($k = 1, 2, 3..$).

Figs. (**5**) and (**6**) illustrate the results obtained for 6CHBT 4-(trans-4′−n-hexylcyclohexyl)-isothiocyanatobenzene with $\Delta\varepsilon \sim 8.0$ and $\Delta n \sim 0.16$ at room temperature.

One can see that plots for 6CHBT (with $\Delta n \sim 0.16$ at room temperature) are in accordance with Eqs. (6) and (7). Moreover, one can observe that plot $I(U)$ for white light (see Fig. **6**) has also a noticeable threshold U_c. Thus the white light may be used for monitoring of the twist deformation in IPS cell (of any cell gap d) when $\gamma = 0$ and polarizers axes adopt orientations at $\alpha = 0$ and $\beta = -\pi/2$.

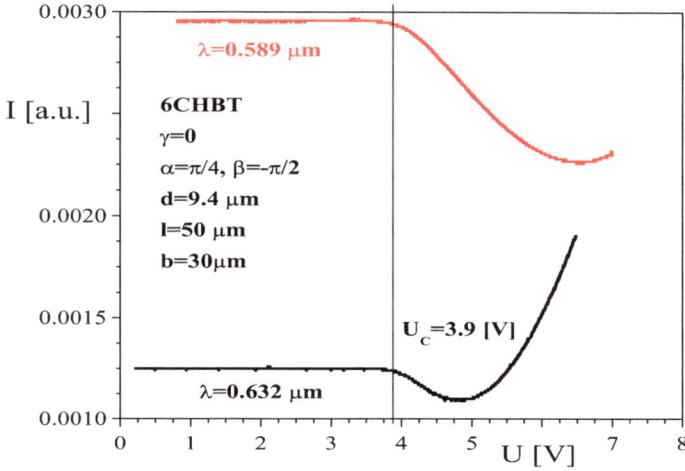

Fig. (5). Twist deformation study. Light intensity I (with $\lambda = 589$ nm and $\lambda = 632$ nm) passing through 6CHBT layer ($\Delta\varepsilon \sim 8.0 > 0$) of thickness $d = 9.4$ μm, *versus* voltage U, applied to electrodes of IPS cell at $\gamma = 0$, $\alpha = \pi/4$, $\beta = -\pi/4$, $l = 50$ μm, $b = 30$ μm.

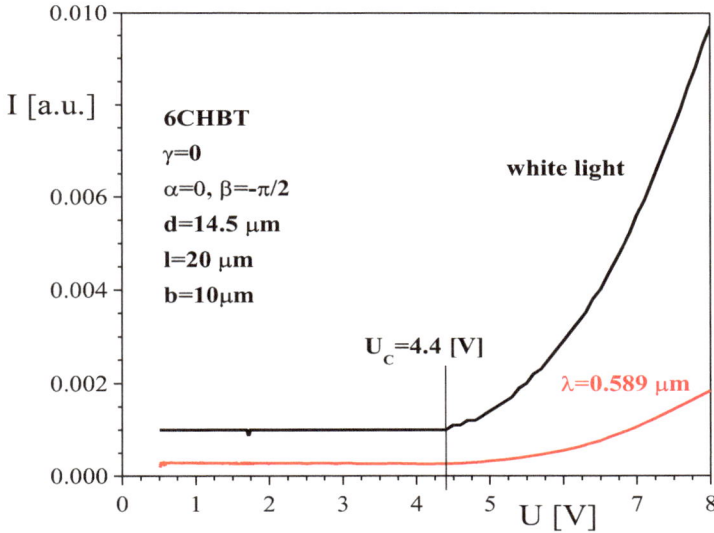

Fig. (6). Twist deformation study. Light intensity I (with $\lambda = 589$ nm and for white light) passing through 6CHBT layer ($\Delta\varepsilon \sim 8.0 > 0$) of thickness $d = 14.5$ μm, versus voltage U applied to electrodes of IPS cell at $\gamma = 0$, $\alpha = 0$, $\beta = -\pi/2$, $l = 20$ μm, $b = 10$ μm.

2.1.2.2. *Optical Monitoring of Twist Deformation Induced by Electric Field E for NNLC ($\Delta\varepsilon$ < 0)*

Let's build an IPS cell (similar to the cell presented in Fig. (**4**)), but with stripes of electrodes perpendicular to the Ox axis. Let's fill this cell (of a cell gap d) with NNLC with $\Delta\varepsilon$ < 0. When rubbing direction is perpendicular to electrodes stripes ($\gamma = 0$) and $U = 0$, the major axis of optical indicatrix of NNLC layer forms an angle $\delta = 0$ and is perpendicular to electrodes.

When voltage $U > U_C$ increases, the twist deformation appears in the NNLC layer. During this deformation, an average optical indicatrix of a deformed NNLC layer changes its shape and spatial orientation of its main axis changes in the same way as it was described previously for PNLC. So, the optical monitoring of twist deformation of the NNLC layer is described by equations (6) and (7) for assumed parameters of alignment (γ) and polarizers geometry (α, β). The test results for a layer of DE with $\Delta\varepsilon$ ~ -1.1 and Δn ~ 0.04 at the room temperature, deformed by an electric field in the IPS cell, are presented in Fig. (**7**).

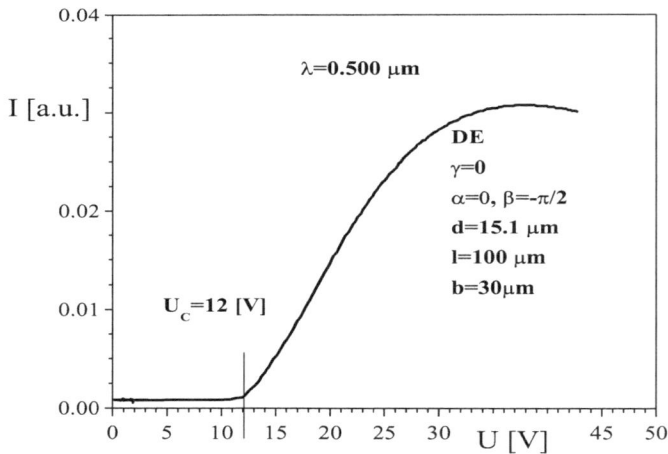

Fig. (7). Twist deformation study. The light intensity I (of $\lambda = 500$ nm) passing through DE layer ($\Delta\varepsilon$ ~ -1.1) of thickness $d = 15.1$ μm as a function of voltage U applied to electrodes of IPS cell when: $\gamma = 0$, $\alpha = 0$, $\beta = -\pi/2$, $l = 100$ μm and $b = 30$ μm.

2.1.3. *Optical Monitoring of Splay (K_{11}) and (K_{33}) Bend Deformations in PNLC ($\Delta\varepsilon$ > 0) and NNLC ($\Delta\varepsilon$ < 0)*

Optical monitoring of splay (K_{11}) and bend (K_{33}) deformations induced by an electric field E in PNLC and NNLC layers was performed in geometries presented in Figs. (**8 – 11**). In all these cases, intensities of light passing through deformed NLC layers are described by the equation:

$$I = 2I_o \tau^2 \sin^2 \frac{\pi d \Delta n}{\lambda}, \tag{7}$$

where $\Delta n(U)$ stands for an effective optical anisotropy of the NLC layer with the director field $\boldsymbol{n}$ deformed by the electric field.

For the given deformation types, when light travels along Oz axis, an effective optical anisotropy $\Delta n(U)$ can be given with formula [6]:

$$\Delta n(U) = \frac{n_o n_e}{\sqrt{n_o^2 \cos^2 \theta(U) + n_e^2 \sin^2 \theta(U)}} - n_o, \tag{8}$$

where $\theta(U)$ is an angle between the director $\boldsymbol{n}$ which was determined in the NNLC layer deformed by the electric field (see Figs. **8-9**). Fig. (**10**) illustrates the optical monitoring of bend deformation in the 6CHBT layer (with $\Delta\varepsilon > 0$) placed in the IPS cell with HT alignment. Thus, we can see that the results presented in Fig. (**10**) confirm the above consideration.

It should be underlined that the curves describing intensity I of light passing through the layers deformed in splay and bend ways by fields $\boldsymbol{E}$, induced by voltage U, applied to electrodes placed on one surface S$_2$ (Figs. **4** and **7**), or applied to two solid electrodes placed on two surfaces S$_1$ and S$_2$ (Fig. **8b**), have the very same nature as the curve shown in Fig. (**10**). Thus, it can be assumed that fields $\boldsymbol{E}$ obtained in given IPS cells were homogeneous enough. In all examples presented in Figs. (**8-9**), due to the application of a very weak rubbing along the direction of axis Oy, the director $\boldsymbol{n}$, tilted at the same angle $\theta(U)$ is located in plane Oyz also.

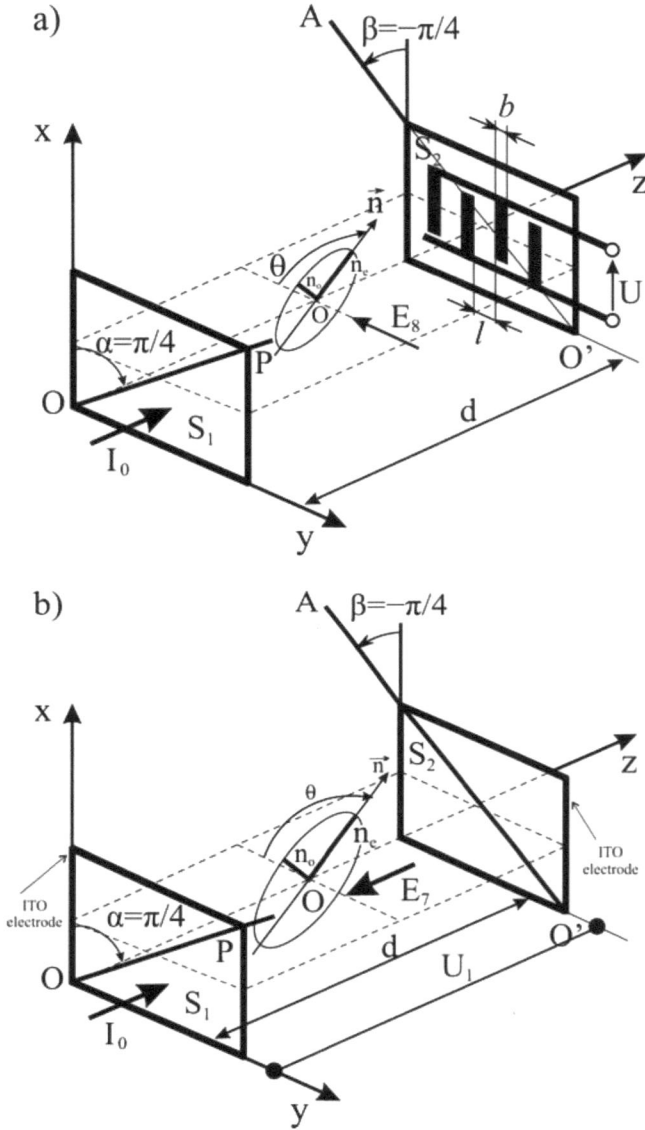

Fig. (8). Bend (K_{33}) deformation study of (a) PNLC layer (with $\Delta\varepsilon > 0$) in IPS cell and HT alignment placed between crossed polarizers. When $U = 0$, $E_8 = 0$ and then θ (0) = $\pi/2$, $\Delta n(U) = 0$. When $U > U_c$ increases, $E_8 > E_{8C}$ increases as well, then θ (U) < $\pi/2$, $\Delta n(U) > 0$; (b) NNLC layer (with $\Delta\varepsilon < 0$) in HT cell placed between crossed polarizers. When $U=0$, $E_7=0$, then θ (0)=$\pi/2$, $\Delta n(U)=0$. When $U > U_c$ increases and $E_7 > E_{7C}$ increases, then θ (U) < $\pi/2$, $\Delta n(U) > 0$.

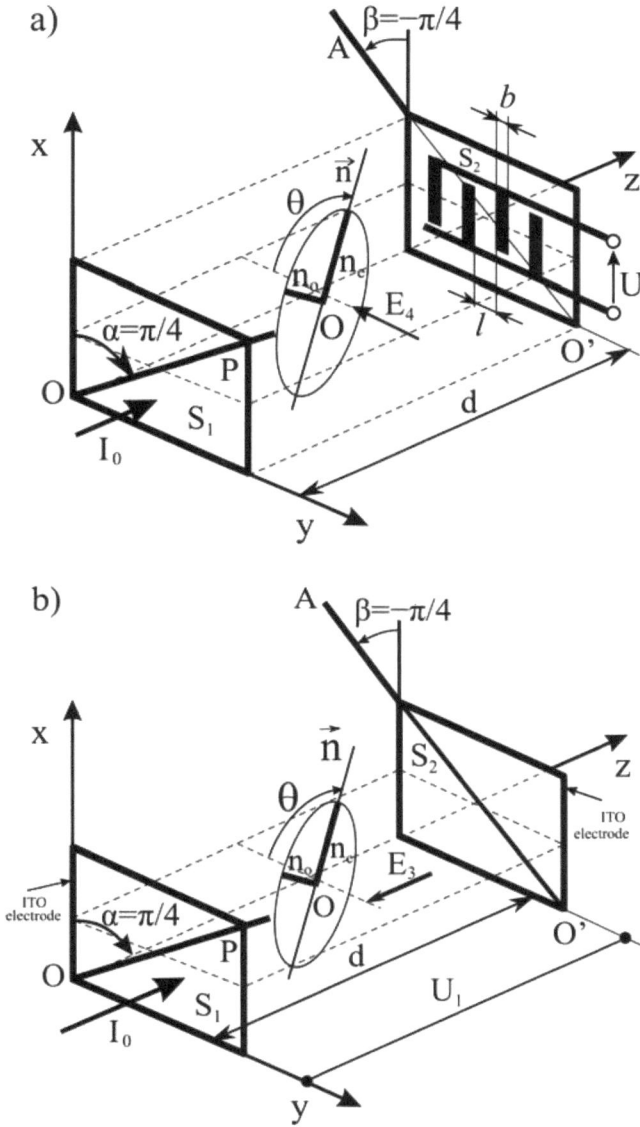

Fig. (9). Splay (K_{11}) deformation study of: (a) NNLC layer (with $\Delta\varepsilon < 0$) in IPS cell and HG alignment placed between crossed polarizers. At $U = 0$ and $E_4 = 0$, one observes $\theta(0) = 0$ and $\Delta n(U) = n_e\text{-}n_o$. When $U > U_C$ increases and $E_4 > E_{4C}$ increases, then $\theta(U) > 0$, $\Delta n(U) < n_e\text{-}n_o$; (b) PNLC layer with $\Delta\varepsilon > 0$ in HG cell placed between crossed polarizers. At $U{=}0$ and $E_4 = 0$, one observes $\theta(0) = 0$ and $\Delta n(U) = n_e\text{-}n_o$. When $U > U_C$ increases and $E_3 > E_{3C}$ increases, then $\theta(U) > 0$, $\Delta n(U) < n_e\text{-}n_o$.

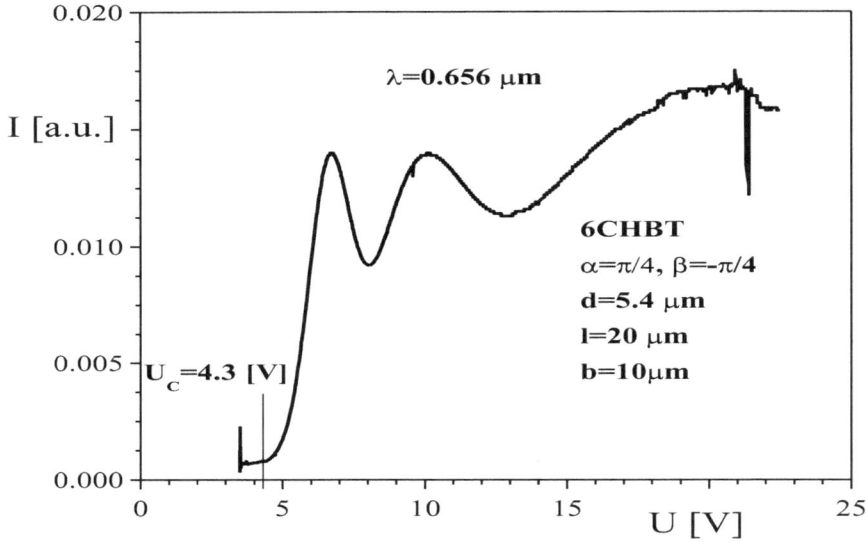

Fig. (10). Bend (K_{33}) deformation study. The light intensity I (at λ = 656 nm) passing through HT aligned 6CHBT slab (d = 14.5 μm, $\Delta\varepsilon \sim$ 8.0) as a function of voltage U applied to IPS electrodes at α = π/4, β = -π/4, l = 20 μm, b = 10 μm.

2.2. On Induction of Homogeneous Electric Fields E in IPS Cells

Knowing values of threshold (critical) voltages U_C as well as l and d parameters of measuring cells, one can calculate critical electric fields $E_C = U_C/l$ in IPS cells and $E_C = U_C/d$ in cells with two solid electrodes for which the tested deformations start.

Table 1. Results of K_{22} assessment for 6CHBT ($\Delta\varepsilon$ = 7.98) in different types of IPS cells at a temperature of 25°C. The intensity of the monitored light I was measured at the aperture $a \gg (l+b)$ under a polarizing microscope.

b [μm]	100			30			10			10		
l [μm]	100			50			40			20		
d [μm]	15.4	9.8	5.0	14.9	9.4	5.3	15.0	9.9	5.1	14.5	10.1	4.8
U_C [V]	5.1	5.4	6.1	4.8	3.9	3.4	7.3	6.7	5.5	4.4	3.5	3.0
K_{22} [pN]	4.41	2.00	0.67	14.6	3.84	0.93	53.2	19.8	3.57	72.8	22.3	3.71

Knowing $\Delta\varepsilon$ (from the dielectric measurements of tested NLC) on the basis of equations (1) and (2), one can easily determine all three elastic constants K_{ii} for this liquid crystal by using the electric field E only. Some results of determining

K_{ii} in different IPS cells (with various d, b and l parameters) for 6CHBT, 5CB as well as for DE are presented in Tables **1** and **2**.

Table 2. Results of K_{ii} measurements for 6CHBT ($\Delta\varepsilon$ = 7.98), 5CB ($\Delta\varepsilon$ = 13.81) and DE ($\Delta\varepsilon$ = -1.03) in different types of IPS cells at 25°C. The intensity of the monitored light I was measured at the aperture of $a \gg (l+b)$ of a polarizing microscope.

	6CHBT	5CB	DE
l [μm]	20	20	20
b [μm]	10	10	10
d [μm]	4.8	5.2	5.0
$\Delta\varepsilon$	8.0, 7.98 [16]	13.8, 13.81 [16]; 3.59 [17]	-1.1 -1.03 [16]
Δn	0.151 [16]	0.184 [16]	0.082 [16]
K_{11} [pN]	6.2, 6.5 [16], 6.71 [18], 7.91 [19]	6.6, 6.1 [16], 6.4 [21], 6.8 [17]	5.4, 5.5 [16]
K_{22} [pN]	3.7, 2.93 [18], 3.64-3.70 [22]	4.1, 4.7 [10]	4.4, 3.7 [10]
K_{33} [pN]	10.0, 9.5 [16], 7.38 [18],5 [20]	9.7, 8.5 [16], 9.85 [23],10.0 [20]	8.4, 8.8 [16]

The analysis of results gathered in Tables **1** and **2** show that results obtained by using IPS cells with different parameters d, b and l, for the same NLC, differ significantly. That is, 6CHBT was observed for which the monitoring process is presented in Figs. (**5**) and (**6**) using a series of cells:

- IPS1 (Fig. **5**): d = 9.4 μm, l = 50 μm, b = 30 μm, U_C = 3.9 V, K_{22} = 3.84 pN,
- IPS2 (Table **1**): d = 5.1 μm, l = 40 μm, b = 10 μm, U_C = 5.5 V, K_{22} = 3.57 pN,
- IPS3 (Fig. **6**): d = 14.5 μm, l = 20 μm, b = 10 μm, U_C = 4.4 V, K_{22} = 72.76 pN.

The value of K_{22} = 3.57 pN (for d = 5.1 μm, l = 40 μm) is close to values determined by using an electric field [10, 24] while constant K_{22} = 72.76 pN is definitely too high for twist deformation in any NCL. The above statement proves that the deforming electric field induced by interdigitating electrodes (b = 10 μm and l = 40 μm) in IPS2 cell with cell gap d = 5 μm is sufficiently homogeneous.

Considering the above and the results of computer calculations which recommended IPS cells and in which the condition $d/(b+l)$ = 0.1 was fulfilled, it was ultimately decided to design and manufacture Hybrid IPS cells (HIPS) based on the following IPS cell parameters: d = 5 μm, l = 40 μm, b = 10 μm.

2.3. Hybrid Measuring Cells with IPS Electrodes (HIPS)

In order to determine three constants (K_{ii}) by using a single measuring cell, two types of HIPS were designed and manufactured at the MUT laboratory [25]:

- 5HIPS40P (d = 5 μm, l = 40 μm, b = 10 μm) for measurements of K_{ii} in PNLC with $\Delta\varepsilon > 0$,
- 5HIPS20N (d = 5 μm, l = 40 μm, b = 10 μm) for measurements of K_{ii} in NNLC with $\Delta\varepsilon < 0$.

2.3.1. Hybrid Measuring Cells 5HIPS40P and 5HIPS40N

Each hybrid cell type 5HIPS40P or 5HIPS40N consists of three separated parts: T, S, and B (as if made of three small measuring cells) placed on the same, common, float glass substrate of a thickness 0.7 mm and refractive index n = 1.54 (see Figs. **11-14**). An active area of each segment (where the electric field should be homogeneous) is 7 x 8 mm. The transparent electrodes of suitable shape were etched in a photolithographic process in ITO layers of specific resistance $\rho < 10$ $\Omega/\square$).

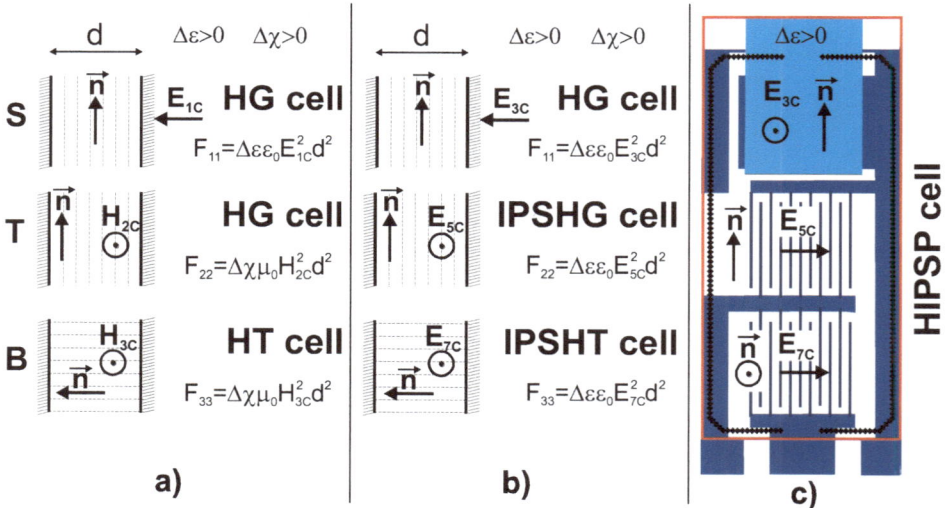

Fig. (11). Geometries of E and H fields and director n orientation directions of PNLC ($\Delta\varepsilon > 0$) while determining splay (K_{11}), twist (K_{22}) and bend (K_{33}) constants at: a) E and H fields applied in three cells with solid electrodes - two HG and HT, b) field E applied in three cells - HG, IPSHG and IPSHT, c) field E applied in one hybrid cell HIPSP only.

Construction schemes and the idea of obtaining homogeneous electric fields with E_3, E_5 and E_7 in 5HIPS40P cell and E_4, E_6 and E_8 in 5HIPS40N cell are presented in Figs. (**11-14**), respectively. In order to obtain a relevant HG orientation of the director n, electrodes were covered with aligning layers of rubbed polyimide NISSAN SE-1199 (TBA ~ 1-2°). Orientation HT was obtained by applying layers of NISSAN SE-1211 (TBA ~ $\pi/2$) polyimide or by evaporation of SiO_2 insulating layers at suitable places. In order to obtain the required thickness d of measuring cells, a spacer of a diameter $d = 5.0$ μm was added to the sealing material and spread on electrodes (2-4 spacers/mm^2). The final thickness of the cells obtained after sealing was measured with an interference meter having an accuracy of ±0.02 μm.

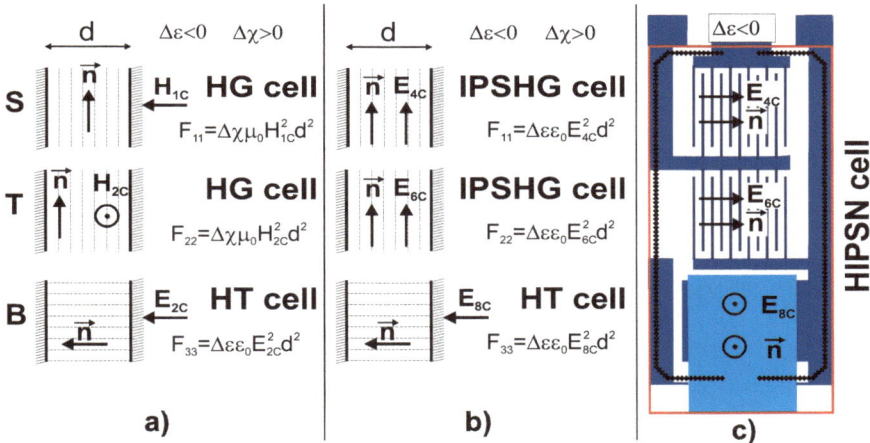

Fig. (12). Geometries of fields E and H and director n orientations directions of NNLC ($\Delta\varepsilon < 0$) applied while determining splay (K_{11}), twist (K_{22}) and bend (K_{33}) constants at: **a)** E and H fields in three cells with solid electrodes - two HG and HT, **b)** E field in three cells - two IPSHG and HT, **c)** E field in hybrid cell HIPSN only.

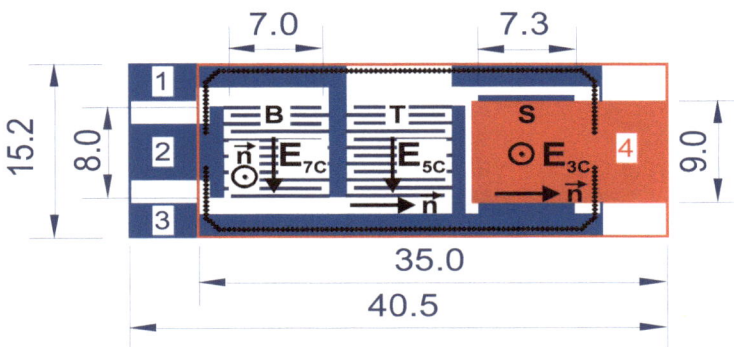

Fig. (13). Conceptual design of HIPSP cell for measuring three K_{ii} constants of PNLC ($\Delta\varepsilon > 0$). In part B of hybrid measuring cell for K_{33} measurement, the polyimide SE-1211 (TBA ~ $\pi/2$, rubbed) for HT orientation was used. In part T for K_{22} measuring and part S for K_{11} measuring polyimide for planar orientation, SE-130 (TBA ~ 1-2°) was applied. Rubbing direction in parts T and S is parallel to IPS electrodes stripes.

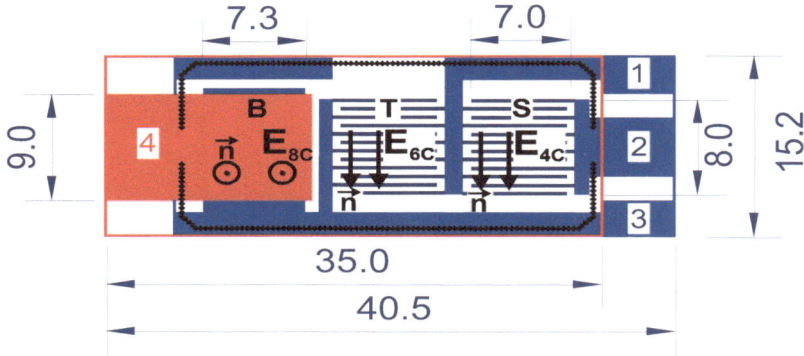

Fig. (14). Conceptual design of HIPSN cell for measuring three K_{ii} constants of NNLC ($\Delta\varepsilon < 0$). In part B of measuring hybrid cell used for K_{33} measuring, the SE-1211 (TBA ~ $\pi/2$) polyimide for HT orientation was used. In part T of measuring hybrid cell for K_{22} measuring, the polyimide SE-130 (TBA~1-2°) for HG orientation, rubbed in the direction perpendicular to IPS electrodes stripes, was applied. In part S of the hybrid cell for K_{11} measurement, the polyimide layer SE-130 (TBA~1-2°) for HG orientation (rubbed in the direction perpendicular to IPS electrodes) was applied.

2.3.2. Evaluation of Hybrid Measuring Cells 5HIPS40P and 5HIPS40N

Hybrid HIPSP and HIPSN cells were tested by using two NCLs, well-known and well described in the literature: 6CHBT ($\Delta\varepsilon > 0$) and DE ($\Delta\varepsilon < 0$). After filling these cells with LCs under study, each part (B, T and S) of a given hybrid cell was optically investigated under a polarizing microscope equipped with a photo-detector. Due to the magnification of microscope (M = 65x), the registered changes of monitoring light intensity I, caused by the change of voltage U applied were measured (at an aperture of a = 2000 μm >> $l+b$ = 50 μm), which were found to be much bigger than the raster c of IPS electrodes. Thus, K_{ii} in 6CHBT ($\Delta\varepsilon > 0$) was determined with the use of 5.1HIPSP40 cell, whereas elastic constants in DE ($\Delta\varepsilon < 0$) were determined in 4.9HIPSN40 cell. The results of the measurement are shown in Table **3**. The elastic constants K_{ii} of 6CHBT ($\Delta\varepsilon > 0$) measured in 5.1HIPSP40 cell and of $\Delta\varepsilon$ ($\Delta\varepsilon < 0$) obtained in 4.9HIPSN40 cell, at 25°C, are in good agreement with data taken from the literature [4, 10, 24, 26, 27]. The compatibility of obtained results brings us to the conclusion that hybrid 5HIPSP40 and 5HIPSN40 cells can be satisfactorily used for quick laboratory measurements of K_{ii} of NLCs with positive ($\Delta\varepsilon > 0$) as well as with negative ($\Delta\varepsilon < 0$) dielectric anisotropy. Such measurements provide basic information regarding the formulation and optimization processes of new, dedicated liquid crystal mixtures with material parameters required in advance.

Table 3. Elastic constants K_{ii} for 6CHBT ($\Delta\varepsilon > 0$) measured in 5.1HIPSP40 cell and for DE ($\Delta\varepsilon < 0$) measured in 4.9HIPS40N cell at 25°C.

	6CHBT in 5.1HIPSP40	DE in 4.9HIPSN40
l [μm]	40	40
b [μm]	10	10
d [μm]	5.1	4.9
K_{11} [pN]	6.7, (6.2 [10], 6.5 [4], 6.7 [24], 7.91 [26])*	5.8, (5.4 [10], 5.5 [4])*
K_{22} [pN]	3.4, (2.93 [24], 3.64-3.70 [27])*	3.7, (4.4 [10])*
K_{33} [pN]	10.6, (9.5 [4], 7.38 [24], 9.5 [4])*	7.9, (8.4 [10], 8.8 [4])*

()* - values from references.

However, one should take note that such a simple and quick assessment of all three K_{ii} constants obtained from one filling of one hybrid cell is subject to a pronounced error δK_{ii}. Error δK_{ii} is caused mainly by a big error δU_C of critical voltages determination U_C (all voltages of Freedericksz's transitions are not satisfactorily sharp, see Figs. (**5-7 and 10**). In case of K_{22} determining in 6CHBT, at $d \sim 5.0$ μm, $\delta d \sim 0.1$ μm, $l \sim 40.0$ μm, $\delta l \sim 1.0$ μm, $U_C \sim 5.5$ V, $U_C \sim 0.5$ V, the relative error $\delta K_{22}/ K_{22}$ is not bigger than 15%.

CONSENT FOR PUBLICATION

Not applicable.

CONFLICT OF INTEREST

The author(s) confirms that there is no conflict of interest.

ACKNOWLEDGEMENTS

Declared none.

REFERENCES

[1]   F.C. Frank, "On the theory of liquid crystals", *Discuss. Faraday Soc.,* vol. 25, pp. 19-28, 1958.
       [http://dx.doi.org/10.1039/df9582500019]

[2]   P.R. Gerber, and M. Schadt, "On the measurement of elastic constants in nematic liquid crystals; Comparison of different methods", *Z. Naturforsch. C,* vol. 35, no. 10, pp. 1036-1044, 1980.
       [http://dx.doi.org/10.1515/zna-1980-1007]

[3]   J. Kędzierski, M.A. Kojdecki, Z. Raszewski, P. Perkowski, J. Rutkowska, L. Lipińska, and E. Miszczyk, "Determination of nematic liquid crystal material parameters by solving inverse problems for different planar cells", *Proc. SPIE,* vol. 4759, pp. 307-312, 2002.
       [http://dx.doi.org/10.1117/12.472167]

[4]   J. Kędzierski, M.A. Kojdecki, Z. Raszewski, J. Zieliński, and L. Lipińska, "Determination of anchoring energy, diamagnetic susceptibility anisotropy, and elasticity of some nematics by means of

semiempirical method of self-consistent director field", *Proc. SPIE,* vol. 6023, pp. 26-40, 2005.
[http://dx.doi.org/10.1117/12.648167]

[5]   J. Kędzierski, Z. Raszewski, M.A. Kojdecki, J. Zieliński, E. Miszczyk, and L. Lipińska, "Optical method for determining anisotropy of diamagnetic susceptibility of nematics and polar anchoring energy coefficient of nematics-substrate systems by using a cell of varying thickness", *Opto-Electron. Rev.,* vol. 12, no. 3, pp. 299-303, 2004.

[6]   J. Kędzierski, M.A. Kojdecki, Z. Raszewski, J. Rutkowska, W. Piecek, P. Perkowski, J. Zieliński, and E. Miszczyk, "Study of anchoring characteristics and splay-bend elastic constant based on experiments with non-twisted nematic liquid crystal cells", *Opto-Electron. Rev.,* vol. 16, no. 4, pp. 390-394, 2008.
[http://dx.doi.org/10.2478/s11772-008-0046-8]

[7]   P. Perkowski, D. Łada, K. Ogrodnik, J. Rutkowska, W. Piecek, and Z. Raszewski, "Technical aspects of dielectric spectroscopy measurements of liquid crystals", *Opto-Electron. Rev.,* vol. 16, no. 3, pp. 271-276, 2008.
[http://dx.doi.org/10.2478/s11772-008-0008-1]

[8]   P. Perkowski, "Dielectric spectroscopy of liquid crystals. Theoretical model of ITO electrodes influence on dielectric measurements", *Opto-Electron. Rev.,* vol. 17, no. 2, p. 180, 2009.
[http://dx.doi.org/10.2478/s11772-008-0062-8]

[9]   Z. Raszewski, "Measurement of permittivity of liquid-crystalline substances", *Electron. Technol.,* vol. 20, p. 99, 1987.

[10]   J. Kędzierski, Z. Raszewski, E. Nowinowski-Kruszelnicki, M.A. Kojdecki, W. Piecek, P. Perkowski, and E. Miszczyk, "Composite method for measurement of splay and bend nematic constants by use of single special in-plane switched cell", *Mol. Cryst. Liq. Cryst. (Phila. Pa.),* vol. 544, pp. 57-68, 2011.
[http://dx.doi.org/10.1080/15421406.2011.569273]

[11]   E. Nowinowski-Kruszelnicki, E. Walczak, A. Kieżun, and L.R. Jaroszewicz, "Light transmission loss in liquid crystal waveguides", *Proc. SPIE,* vol. 3318, pp. 410-413, 1998.
[http://dx.doi.org/10.1117/12.300014]

[12]   M. Born, and E. Wolf, *Principles of Optics* (7th ed., revised). Cambridge: Cambridge University Press, 2003.

[13]   M. Pluta, *"Advanced Light Microscopy"*, Specialized Methods, Elsevier: Amsterdam-Oxfor-New York-Tokyo, PWN-Polish Scientific Publishers, Warszawa, 1989.

[14]   M. Pluta, "Simplified polanret system for microscopy", *Appl. Opt.,* vol. 28, no. 8, pp. 1453-1466, 1989.
[http://dx.doi.org/10.1364/AO.28.001453] [PMID: 20548681]

[15]   P. Perkowski, Z. Raszewski, J. Kędzierski, W. Piecek, J. Rutkowska, J. Zieliński, R. Dąbrowski, and W.J. Drzewiński, "Computer calculation for refractive indices in smectic phases", *Ferroelectrics,* vol. 309, pp. 55-61, 2004.
[http://dx.doi.org/10.1080/00150190490509917]

[16]   A. Walczak, E. Nowinowski-Kruszelnicki, L. Jaroszewicz, and R. Wal, "Edge detection with liquid crystal polarizing filter", *Mol. Cryst. Liq. Cryst. (Phila. Pa.),* vol. 413, pp. 407-415, 2004.
[http://dx.doi.org/10.1080/15421400490438924]

[17]   C. Tyszkiewicz, T. Pustelny, and E. Nowinowski-Kruszelnicki, "Investigation of a ferronematic cell influenced by a magnetic field", *J. Phys.,* vol. 137, pp. 161-164, 2006.

[18]   S. Ertman, T.R. Woliński, D. Pysz, R. Buczyński, E. Nowinowski-Kruszelnicki, and R. Dąbrowski, "Tunable broadband in-fiber polarizer based on photonic liquid crystal fiber", *Mol. Cryst. Liq. Cryst. (Phila. Pa.),* vol. 502, pp. 87-98, 2009.
[http://dx.doi.org/10.1080/15421400902815779]

[19]   S. Kłosowicz, and E. Nowinowski-Kruszelnicki, "PDLC systems in elliptical capillaries", *Mol. Cryst. Liq. Cryst. (Phila. Pa.),* vol. 375, pp. 205-214, 2002.

[http://dx.doi.org/10.1080/713738338]

[20] A. Walczak, E. Nowinowski-Kruszelnicki, L. Jaroszewicz, and P. Marciniak, "Tuned liquid crystalline interferometer analysis by means of generalised Berreman matrix", *Opto-Electron. Rev.,* vol. 10, no. 1, pp. 69-73, 2002.

[21] A. Walczak, and E. Nowinowski-Kruszelnicki, "Polarization sensitive liquid crystal filter for polarization difference imaging", *Proc. SPIE,* vol. 5947, pp. 76-82, 2005. [http://dx.doi.org/10.1117/12.622801]

[22] J. Zielinski, E. Nowinowski-Kruszelnicki, R. Dąbrowski, and J. Żmija, "Twisted-nematic effect in the mixtures of p-alkylbenzoates of p-cyanophenol", *Electron. Technol.,* vol. 11, no. 3, p. 122, 1978.

[23] M. Sutkowski, P. Garbat, E. Nowinowski-Kruszelnicki, A. Walczak, J. Parka, and J. Woźnicki, "Polarization difference image analysis with LC filter", *Opto-Electron. Rev.,* vol. 17, no. 1, pp. 53-58, 2009. [http://dx.doi.org/10.2478/s11772-008-0035-y]

[24] P. Buchecker, and M. Schadt, "Synthesis, physical properties and structural relations of new, end-chain substituted nematic liquid crystals", *Mol. Cryst. Liq. Cryst. (Phila. Pa.),* vol. 149, pp. 359-373, 1987. [http://dx.doi.org/10.1080/00268948708082991]

[25] E. Nowinowski-Kruszelnicki, J. Kędzierski, and Z. Raszewski, "Measurement of Elastic Constants of Nematic Liquid Crystal With Use of Hybrid In-Plane-Switched Cell", *Opto-Electron. Rev,* vol. 20, p. 255 , 2012.

[26] Z. Raszewski, J. Kędzierski, J. Rutkowska, J. Zieliński, J. Żmija, R. Dąbrowski, and T. Opara, "Dielectric investigation of the diamagnetic anisotropy and elasticity of 4-trans-4'-n-he-syl-cycloheksyl-isothiocyanatobenzene (6CHBT)", *Liq. Cryst.,* vol. 14, pp. 1959-1966, 1993. [http://dx.doi.org/10.1080/02678299308027732]

[27] J.W. Baran, Z. Raszewski, J. Kędzierski, and J. Rutkowska, "Some physical properties of mesogenic 4-(trans-4'-n-alkylcyclohexyl)isothiocyanatobenzenes", *Mol. Cryst. Liq. Cryst. (Phila. Pa.),* vol. 123, pp. 237-245, 1985. [http://dx.doi.org/10.1080/00268948508074781]

CHAPTER 3

Nematic Mixtures for Liquid Crystal Light Modulators

Przemysław Kula[*]

Military University of Technology, Warsaw, Poland

Abstract: In the last few years, the following Multicomponent Nematic Liquid Crystalline (MNLC) mixtures of medium and high optical anisotropy Δn have been developed in Military University of Technology (MUT) in Warsaw:

• W1898 (composed mainly of two and three alkylcyclohexylbenzene and alkylbicyclo-hexylbenzene isothiocyanates),

• W1820 (composed mainly of fluoro-substituted alkylbiphenylisothiocyanates),

• W1852 (composed mainly of fluoro-substituted isothiocyanates, alkylbiphenyls, alkylcyclo-hexylbiphenyls and alkylbicyclocheksylobiphenyls),

• W1825 and W1791 (composed mainly of fluoro-substituted isothiocyanates, alkyltolanes and alkylphenyltolanes),

• W1865 (composed mainly of fluoro-substituted isothiocyanates, alkylphenyls and alkylbiphenyltolanes),

• W1115 (composed mainly of cyclohexanolcyanobenzenes, fluorobenzenes and hydrocarbons),

• W1795B (composed mainly of fluoro-terphenyls).

In this chapter, the material characteristics of the perpendicular $\varepsilon_\perp(T)$ and parallel $\varepsilon_{//}(T)$ components of the permittivity tensors ε (T) at the frequency range of $f \in$ [10 Hz, 10 MHz]; bulk viscosity $\Gamma(T)$; rotational viscosity $\gamma(T)$, splay $K_{11}(T)$, twist $K_{22}(T)$, bend $K_{33}(T)$ and $K_{TN}(T)$ for transition from twisted to the nontwisted structure; ordinary $n_o(\lambda)$ and extraordinary $n_e(\lambda)$ refractive indices for wavelength $\lambda \in$ [0.3μm, 1.6 μm] are described and discussed in the terms of applicability of MNLCs to Liquid Crystal electro-optical Devices (LCEOD).

Keywords: Birefringence fringes, Bulk viscosity, Homogeneous, Homeotropic

---
[*] **Corresponding author Przemysław Kula:** Military University of Technology, Faculty of Advanced Technologies and Chemistry, Warsaw, Poland; E-mail: przemyslaw.kula@wat.edu.pl

Leszek R. Jaroszewicz (Ed.)

and twisted alignment, Liquid crystal cell, Light modulators for naval, Negative nematic liquid crystals, Optical anisotropy, Positive nematic liquid crystals, Refractive indices, Rotational viscosity, Splay, Switching times, Twist.

1. INTRODUCTION

Unsubstituted cyclohexyl-benzene isothiocyanates and dicyclohexyl-benzene, fluoro-terphenyls, tolanes, as well as isothiocyanates fenylotolanes, cyclohexa-nolcyanobenzenes, fluorobenzenes, and hydrocarbons are extremely useful components of Multicomponent Nematic Liquid Crystalline (MNLC) mixtures of medium and high optical anisotropy Δn [1]. These compounds, often exhibiting an enantiotropic nematic phase, are very attractive components of highly specialized liquid crystalline mixtures with relatively low viscosity γ and a relatively high permittivity anisotropy $\Delta \varepsilon$. By varying individual proportions of components in MNLC (*e.g.*, W1115 is a 16-component mixture) one can modify in a fairly broad range not only their electric ($\varepsilon_\perp$, $\Delta \varepsilon$), optical (Δn), viscous (γ) and elastic parameters (K_{ii}) but also a temperature range of the nematic liquid crystal phase [2, 3].

In this chapter, the results of studies of physical properties of eight liquid crystalline mixtures with medium and high-birefringence have been compiled. They can be applied as working mixtures for liquid crystal light modulators. Examples of elaborated working mixtures are:

- W1898 (composed mainly of two and three alkylcyclohexylbenzene and alkylbicyclohexylbenzene isothiocyanates),
- W1820 (composed mainly of fluoro-substituted alkylbiphenyl- isothiocyanates),
- W1852 (composed mainly of fluoro-substituted isothiocyanates, alkylbiphenyls, alkylcyclohexylbiphenyls and alkylbicycloheksylobiphenyls),
- W1825 and W1791 (composed mainly of fluoro-substituted isothiocyanates, alkyltolanes and alkylphenyltolanes),
- W1865 (composed mainly of fluoro-substituted isothiocyanates, alkylphenyls and alkylbiphenyltolanes),
- W1115 (composed mainly of cyclohexanolcyanobenzenes, fluorobenzenes and hydrocarbons),
- W1795B (composed mainly of fluoro-terphenyls).

2. CELLS FOR MEASUREMENTS OF LIQUID CRYSTALS PROPERTIES

In order to determine the mesogenic behavior and dielectric, refractometric, spectral and electro-optical characteristics of tested MNLC mixtures three types of cells for measurements of liquid crystals properties were designed and manufactured at the laboratory of Crystals Physics and Technology at the Military University of Technology (MUT). The cells have been made under WAT1 standard (with an active area between electrodes of a size 5.08×5.08 mm) and under WAT2 standard (with an active area of 12.7×12.7 mm). The inner surface of cells ensures homogeneous (HG), homeotropic (HT) or twisted (TN) alignment of the liquid crystalline structure. The layout of these cells is presented in Fig. (**1**). Both substrates of measuring cells were made of high-quality float glass plates usually applied for the manufacturing of liquid crystal displays. Substrates made of float glass of thickness 0.7 mm were covered with a transparent, conductive layer of Indium Tin Oxide (ITO *i.e.* $In_{2-x}Sn_xO_3$) of sheet resistance of 10, 70, 100 or 500 $\Omega/\square$. The proper shape of electrodes was formed upon a wet etching process. At the process of fabrication of cells for the dielectric measurements, a thin, opaque, conductive layer of gold (Au) with a chromium (Cr) adhesive layer was deposited over the glass surface. The sheet resistance of such electrodes was smaller than 0.01 $\Omega/\square$.

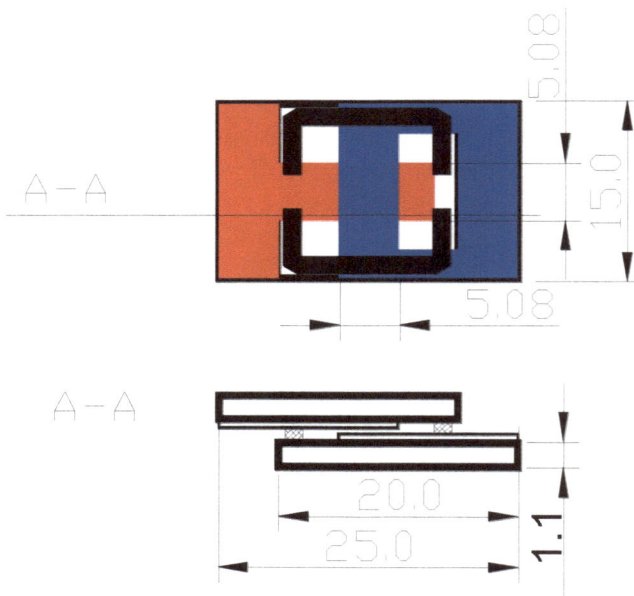

Fig. (1). Layout of a cell manufactured under WAT1 standard. Red and blue colors denote conductive layers on the bottom and top substrates, respectively.

In order to achieve planar orientation of the molecular director ***n*** within cells of HG and TN type electrodes were spin-coated with polyimide alignment layers provided by NISSAN Chemical Industries. For homogeneous HG orientation, the SE-130 polyimide was used while polyimide SE-1211 was applied to achieve homeotropic HT orientation. After spin-coating of pre-polymer, substrate glasses were subsequently dried and baked for cyclization. Subsequently, the polyimide polymer-coated surfaces were rubbed unidirectionally with a dedicated cloth (YA-19R by Yoshikawa Chemical Company) in order to achieve surface anisotropy. In some cases, a SiO_2 film was thermally vaporized at a proper angle to the substrate normal, as an inorganic alignment layer. In order to ensure a proper cell gap d, special cylindrical or spheroidal glass spacers of the diameters like 1.6, 2.0, 2.5, 3.0, 5.0, 6.0, 7.0, 8.0 and 10.0 µm were added to the sealing adhesive. Additional spacers were sprayed over the electrode surface in an amount of 2- 4 pcs/mm². The final cells gap d was developed upon curing the glue within a hot-press. The final cell gap d was determined by an interference method.

Following a system of labels was introduced to identify parameters of measuring cells used for the materials studies:

- 1.6HG10 stands for the cell with a cell gap d=1.6 µm, equipped ITO electrodes of specific surface resistance ρ=10 Ω/□ and with HG alignment layer,
- 5.3HT500 stands for a cell with HT alignment, with a cell gap d=5.3 µm and with ITO electrodes of specific surface resistance ρ=500 Ω/□,
- 8.4HTAu stands for a cell with HT alignment, with a cell gap d=8.4 µm and with Au electrodes of specific surface resistance ρ=0.01 Ω/□,
- 6.6TN70 stands for a cell with TN alignment, with a cell gap d=6.6 µm and with ITO electrodes of specific surface resistance ρ=70 Ω/□.

3. NEMATIC LIQUID CRYSTALLINE MIXTURES

Phase transition temperatures of MNLC mixtures: N-T °C -**I** from isotropic (**I**) phase to nematic (**N**) phase and **Cr**-T °C -**N** from nematic to crystal (**Cr**) phase were assessed by using a BIOLAR PI polarizing microscope equipped with a hot stage LINKAM HMSE600 controlled by a TMS92 driving unit. Phase transition temperatures as well as the (translational) viscosity coefficient γ of tested MNLC mixtures were obtained by using an Ostwald's capillary viscometer. The results are presented in Table **1** at the end of this Chapter.

3.1. Dielectric Study of W1898 and W1795B Mixtures

The dielectric measurements of tested MNLC mixtures were carried out by means of LF Impedance Analyzer HP 4192A at the frequencies $f \in$ (10 Hz; 10 MHz) at the temperature domain from 5°C to 100°C. The temperature stabilization during the measurements was better than 0.2 K [4 - 8]. Typical temperature characteristics of the components of real parts of the permittivity tensor (of $\varepsilon'_\parallel(T)$ and $\varepsilon'_\parallel(T)$) of the W1898 (with $\Delta\varepsilon > 0$) are presented in Fig. (**2**). Fig. (**3**) shows the analogous characteristics for the mixture W1795B (with $\Delta\varepsilon < 0$).

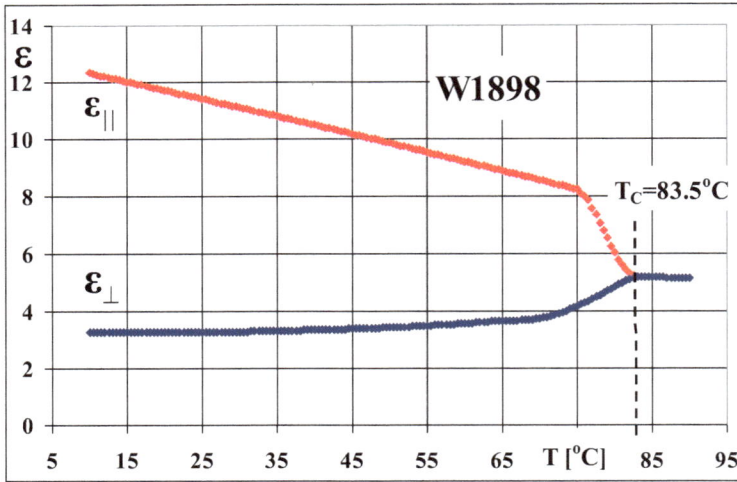

Fig. (2). Temperature characteristics of $\varepsilon'_\parallel(T)$ and $\varepsilon'_\perp(T)$ real parts of the permittivity of the W1898 mixture observed at the frequency $f = 1.0$ kHz, determined in 5.4HGAu and 5.2HTAu cells, respectively.

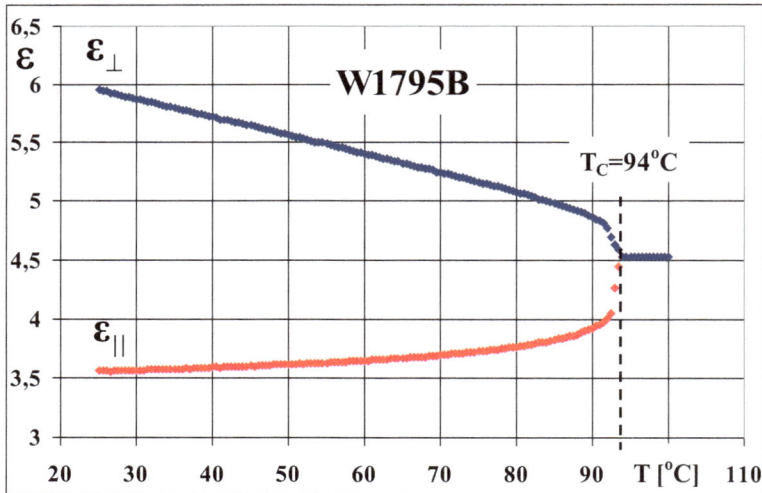

Fig. (3). Temperature characteristics of $\varepsilon'_\parallel(T)$ and $\varepsilon'_\perp(T)$ real parts of W1795B mixture at the frequency $f = 1.0$ kHz, determined in 5.3HGAu and 5.3HTAu cells, respectively.

Dielectric dispersive measurements were carried out at the frequency range from 100 Hz to 10 MHz at various temperatures T. A typical graph of the real $\varepsilon'_\parallel$ and imaginary $\varepsilon_\parallel''$ parts as a function of the frequency f for the W1898 mixture at $T = 25°C$ is shown in Fig. (**4**). Thermal and dispersive characteristics of real parts $\varepsilon_\parallel'$ and $\varepsilon_\perp'$ at the frequency of $f = 1.0$ kHz and imaginary part $\varepsilon_\parallel''(f)$ at 25°C described above were determined for all eight selected liquid crystal mixtures. The results of the studies are discussed in this Chapter and presented in Table **1**.

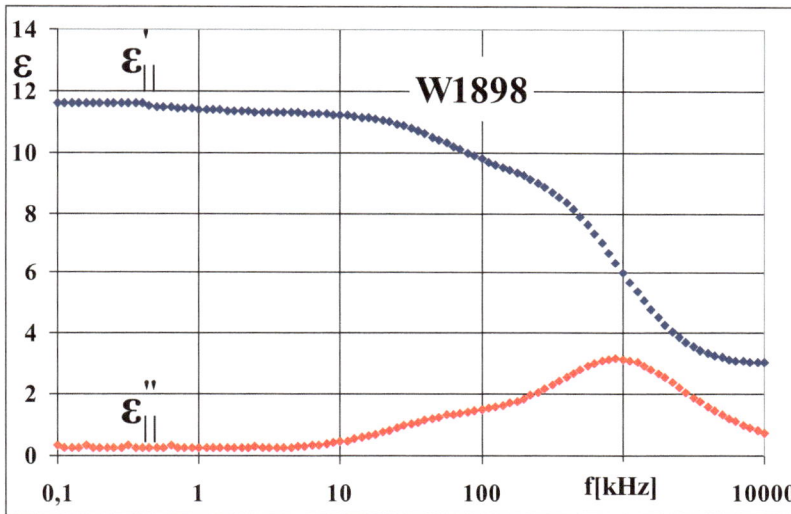

Fig. (4). Dispersion characteristics of a $\varepsilon_\parallel'(f)$ real and $\varepsilon_\parallel''(f)$ imaginary parts of the permittivity of W1898 mixture (at 25°C) obtained using a 5.3HTAu measuring cell.

3.2. Refractometric Study of W1898 Mixture

Temperature characteristics of the refractive indices n_i (T) of the MNLC mixtures at the isotropic phase, as well as ordinary no(T) and extraordinary ne(T)refractive indices at the nematic phase, were measured by means of a set of specially adapted Abbe refractometers. By technical limitations and using this method the refractive indices $n(T)$ smaller than $n < 1.87$ can be determined at the temperatures below 130°C. The temperature stabilization while measurements were better than 0.2 K [4]. The typical thermal characteristics or refractive indices $n_i(T)$, $n_o(T)$ and $n_e(T)$ of mixture W1898 are presented in Fig. (**5**).

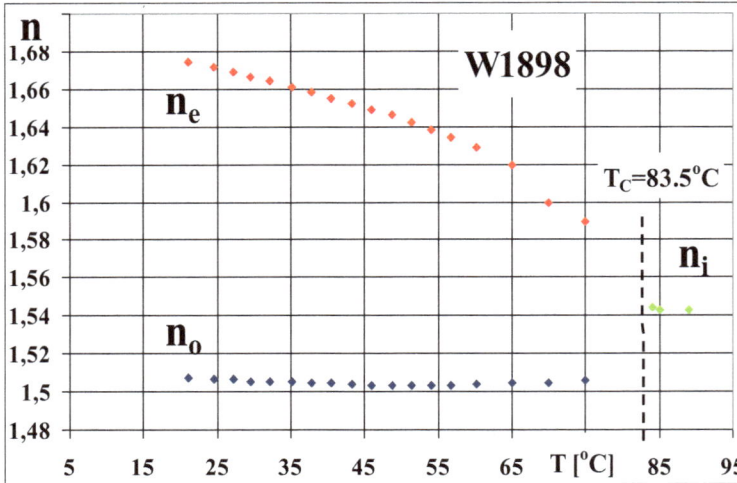

Fig. (5). Temperature characteristics of refractive indices $n_i(T)$, $n_o(T)$ and $n_e(T)$ of W1898 mixture obtained at $\lambda = 0.5893$ µm.

In order to determine refractive indices n being out of range of the Abbe refractometer (for n exceeded 1.87) as well as to determine them at Near Infra-Red (NIR) spectrum (up to $\lambda = 1.5$ µm), an interference method using wedge-shaped cells [8, 9] was applied. Interference fringes were observed at VIS and NIR ranges for a wedge-shaped cell filled with tested MNLC mixtures. The observed distribution of interference fringes is a basis for the determination of values of the refractive indices n_o, n_e as well as their anisotropy $\Delta n = n_e - n_o$. Interference fringes were registered by a CCD camera without NIR filter. Interference fringes observed at the temperature of 25°C at the wavelength of $\lambda = 0.6328$ µm using a wedge-shaped cell filled with the W1898 mixture are presented in Fig. (6).

In order to check the correctness of measurement methods of refractive indices which are described above and as to broaden the spectrum scope (up to $\lambda = 1.5$ µm) of n determination, the additional interference measurements were performed by using a JASCO V670 spectrometer. Birefringence fringes for a cell 7.9HG500 filled with mixture W1898 placed between crossed polarizers are presented in Fig. (7). The dispersion of the optical anisotropy $\Delta n(\lambda_k)$ of MNLC mixtures can be calculated according to formula $\Delta n(\lambda_k) = (2k-1)\lambda_k/(2d)$ [10], where:

k - (interference) order of a fringe, λ_k - spectral position of the centre of k fringe, d - the cell gap.

Fig. (6). Interference fringes observed in a wedge-shaped cell filled with the W1898 mixture at 25°C illuminated with the laser beam of a wavelength $\lambda = 0.6328$ μm: **a**) the empty wedge cell ($n = 1$), interference fringes period $A = 0.0597$ mm, **b**) filled cell observed at the direction of light polarization parallel with the director n ($n = n_e$), fringes period $E = 0.0357$ mm, calculated $n_e = A/E = 1.672$ [9], **c**) filled cell at the direction of light polarization perpendicular to the director n ($n = n_o$), interference fringes period $O = 0.0396$ mm, calculated $n_o = A/O = 1.508$ [9], **d**) filled cell placed between crossed polarizers (director n forms the angles of 45° and -45° with the polarizer's and analyzer's axes respectively), interference fringes period $B = 0.7067$ mm, calculated optical anisotropy $\Delta n = 2A/B = 0.169$ [9].

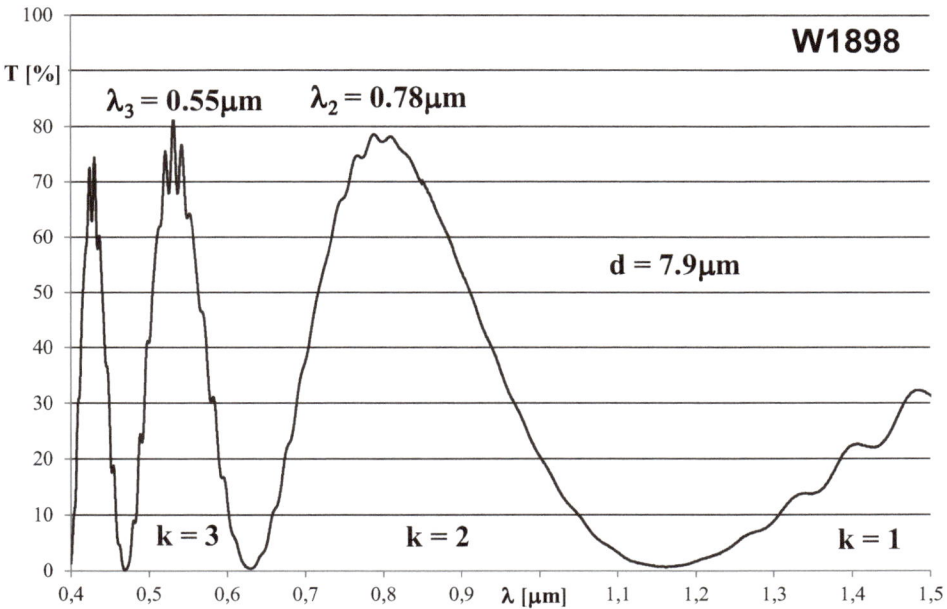

Fig. (7). Interference fringes observed at the temperature of 25°C for 7.9HG500 cell filled with W1898 mixture. Transmission spectrum $T(\lambda)$ was registered by means of JASCO V670 spectrometer.

The use of n_e (λ) and n_o (λ) values measured with the use of the Abbe refractometer as well as from interference measurements with wedge cells and planar HG cells, the dispersion curves of the optical anisotropy $\Delta n(\lambda)$ of MNLC mixtures under study were determined. An example of such a dispersion curve Δn (λ) at the VIS-NIR range is shown in Fig. (**8**). Thermal characteristics of n_e (T) and n_o (T) described above as well as dispersions Δn (λ) within VIS-NIR range were determined for eight selected liquid crystalline mixtures. The results of a study are presented in Table **1**.

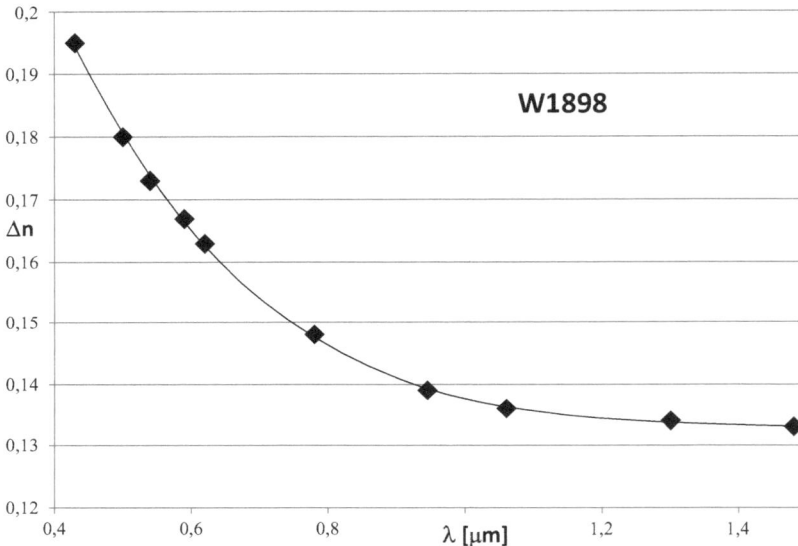

Fig. (8). The dispersion of an optical anisotropy $\Delta n(\lambda)$ for W1898 mixture observed at 25°C. Points come from combined method utilizing Abbe refractometer as well as from the interference measurements.

3.3. Determination of Frank Elastic Constants K_{ii} of W1898 and W1795B Mixtures

By filling a single 5.1HIPSN40 measuring cell (see Chapter 2) a three K_{11}, K_{22} and K_{33} elastic constants of W1795B mixture with a negative permittivity anisotropy $\Delta\varepsilon = -2.5$ (see Fig. **3**) and a significantly high optical anisotropy $\Delta n = 0.23$ were evaluated (see Table **1**). The results of K_{ii} study of W1795B mixture being an example of MNLC mixtures with $\Delta\varepsilon < 0$ are presented in Figs. (**9 - 12**) as well as in Table **1**.

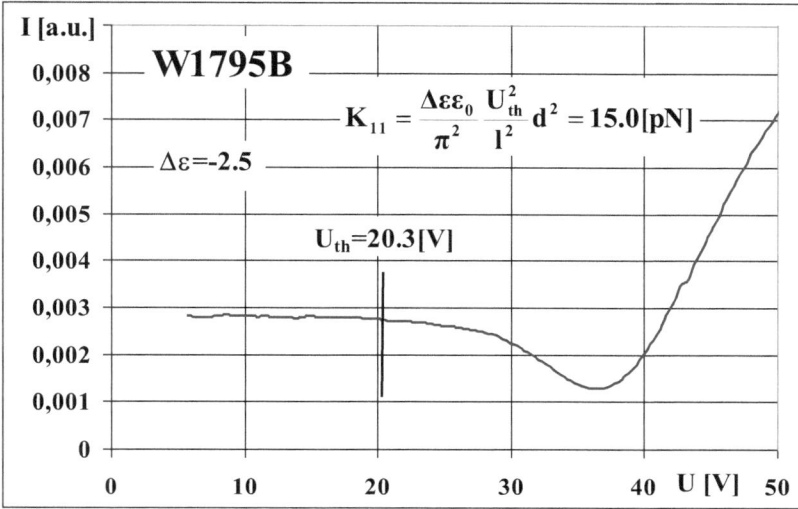

Fig. (9). Splay deformation of the W1795B mixture with $\Delta\varepsilon$ = -2.5 and Δn = 0.23 at 25°C monitored by light (λ = 0.589 μm) in a hybrid 5.1HIPSN40 cell. Voltage U was applied to electrodes 1 and 2 in section S of the cell (see Fig. **14**, Chapter 2).

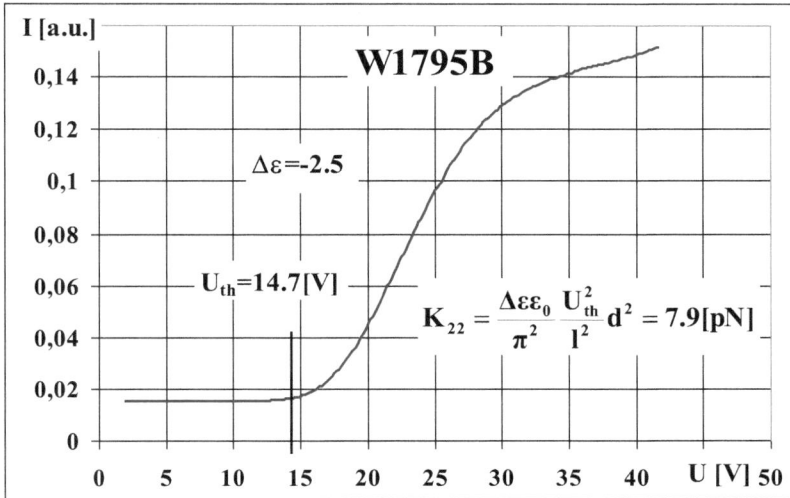

Fig. (10). Twist deformation of the W1795B mixture with $\Delta\varepsilon$ = -2.5 and Δn = 0.23 at 25°C monitored by light (λ = 0.589 μm) in a hybrid 5.1HIPSN40 cell. Voltage U was applied to electrodes 1 and 3 in section T of the cell (see Fig. **14**, Chapter 2).

The determination of the K_{33} bend constant in section B of a hybrid 5HIPSN40 cell (see in Figs. **11** and **12**) for MNLC mixtures (with $\Delta\varepsilon$ < 0) can be carried out by using two methods. The threshold voltages U_{th} of Freedericksz's transition for bend deformation can be determined by both optical (Fig. **11**) and dielectric monitoring (Fig. **12**). It can be noted that thresholds voltages U_{th} are more precise when optical monitoring of the Freedericksz's transition is utilized.

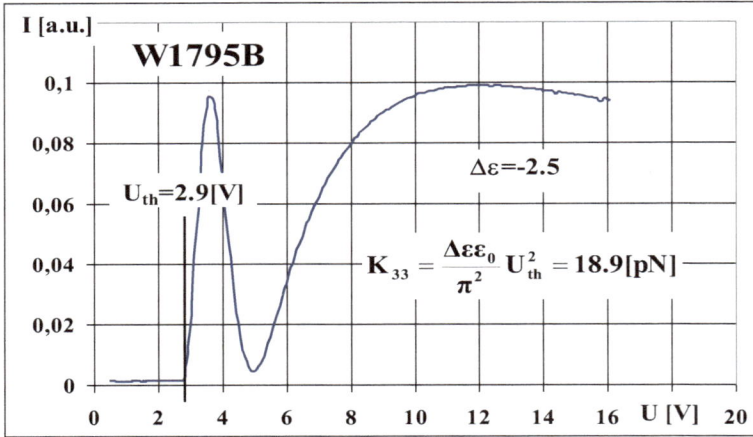

Fig. (11). Bend deformation of the W1795B mixture with $\Delta\varepsilon = -2.5$ and $\Delta n = 0.23$ at 25°C monitored by light ($\lambda = 0.589$ μm) in a 5.1HIPSN40 cell. The voltage U was applied to electrodes 3 and 4 in section B of the cell (see Fig. **14**, Chapter 2).

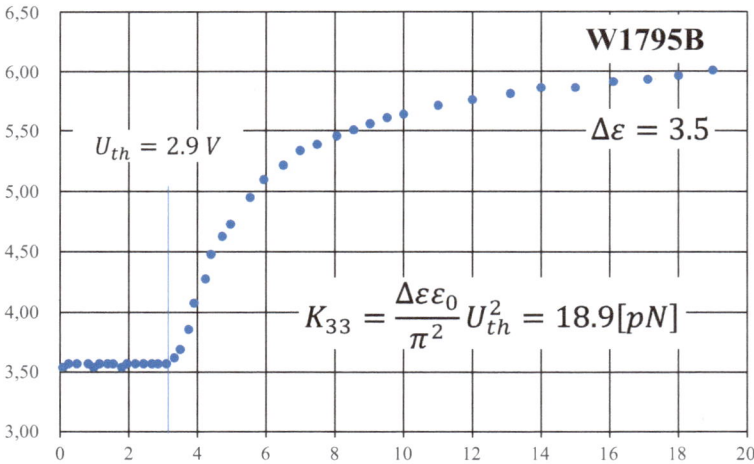

Fig. (12). Bend deformation of the W1795B mixture with $\Delta\varepsilon = -2.5$ and $\Delta n = 0.23$ at 25°C monitored at the dielectric study at the frequency $f = 1.5$ kHz in a hybrid 5.1HIPSN40 cell. The voltage U was applied to electrodes 3 and 4 in section B of the cell (see Fig. **14**, Chapter 2).

The elastic constants K_{ii} of seven, newly developed at MUT, liquid crystalline mixtures (W1898, W1820, W1852, W1825, W1791, W1865, W1115) with $\Delta\varepsilon > 0$ and the high optical anisotropy Δn were determined in a similar way. For the determination of K_{ii} of these mixtures, the hybrid 5HIPSP40 or 5HIPSO20 cells were applied. The results of K_{ii} study for W1898 with $\Delta\varepsilon = 8.1$ and $\Delta n = 0.17$ at 25°C are presented in Figs. (**13 - 16**). The results of studies of all MNCL mixtures are presented in Table **1**.

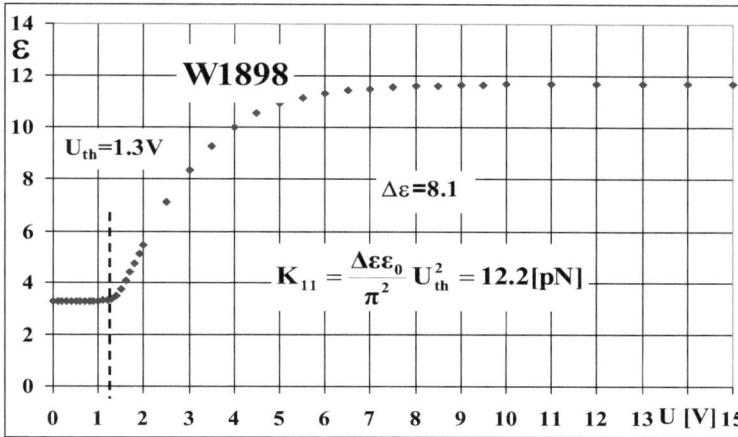

Fig. (13). Splay deformation of the W1898 mixture monitored at the dielectric study, at the frequency $f = 1.5$ kHz by using a hybrid 5.1HIPSN40 cell. Voltage U was applied to electrodes 3 and 4 in section S of the cell (see Fig. **13**, Chapter 2).

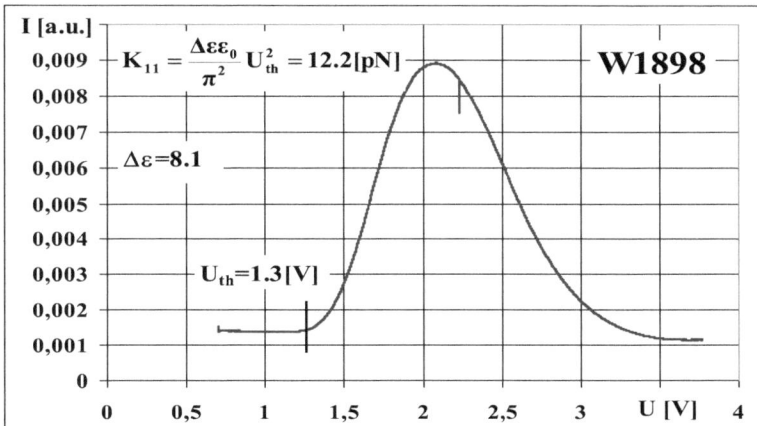

Fig. (14). Splay deformation of the W1898 mixture monitored optically ($\lambda = 0.589$ μm) using a hybrid 5.1HIPSP20 cell. Voltage U was applied to electrodes 3 and 4 in section S of the cell (see Fig. **13**, Chapter 2). The polarization plane of incident light forms the angle $\alpha = \pi/4$ with the director **n**.

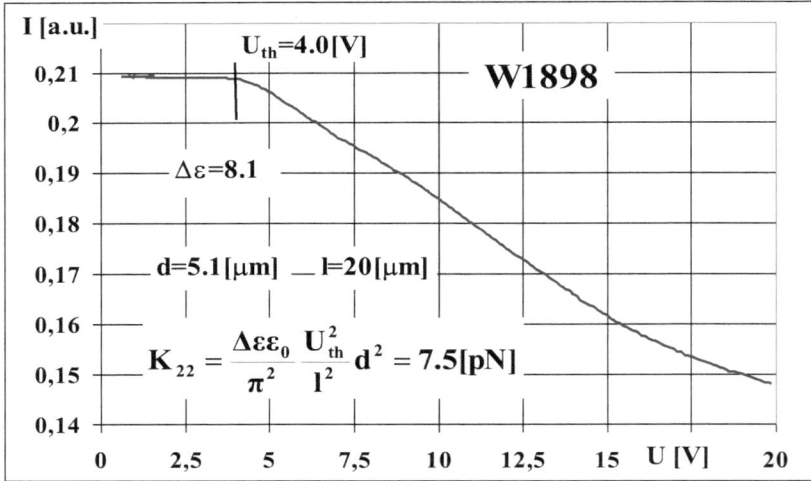

Fig. (15). Twist deformation of the W1898 mixture monitored by light ($\lambda = 0.589$ μm) in a hybrid 5.1HIPSP20 cell ($d = 5.1$ μm, $b = 10$ μm, $l = 20$ μm). Voltage U was applied to electrodes 1 and 3 in section T of the cell (see Fig. **14**, Chapter 2). Polarization plane of an incident light forms the angle $\alpha = \pi/4$ with director $\boldsymbol{n}$.

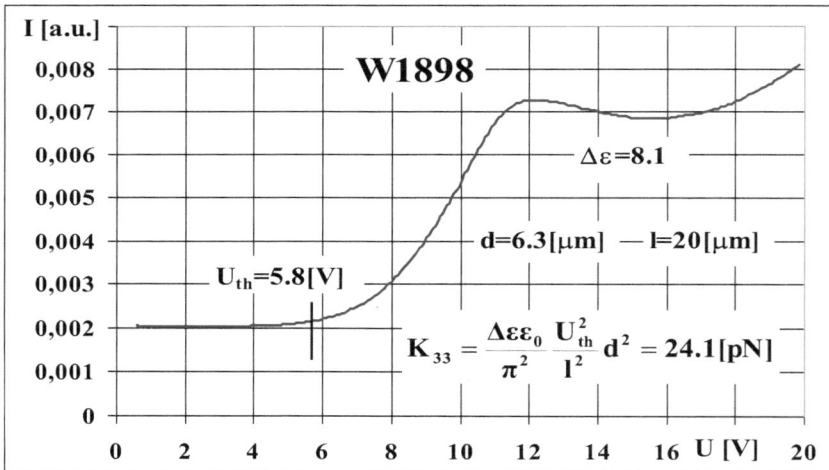

Fig. (16). Bend deformation of the W1898 mixture monitored by light ($\lambda = 0.589$ μm) in a hybrid 5.1HIPSP20 cell ($d = 5.1$ μm, $b = 10$ μm, $l = 20$ μm). Voltage U was applied to electrodes 1 and 2 in section B of the cell (see Fig. **14**, Chapter 2). Section B of a hybrid 5.1HIPSP20 cell with W1898 was placed between crossed polarizers.

3.4. Determination of K_{TN} Reduced Elastic Constant of W1898 Mixture

The obtained values of K_{ii} constants for seven MNCL mixtures with $\Delta\varepsilon > 0$ were verified by determination of a reduced elastic constant K_{TN} at an observation of Twisted Nematic (TN) electro-optical effect in plain twisted TN cells (angle of a layer, twisting was $\xi = \pi/2$). The results of K_{TN} determination for W1898 in the 3.1TN70 cell are presented in Figs. (**17, 18**) and in Table **1**. K_{TN} constants gathered in Table **1** for other MNCL mixtures with $\Delta\varepsilon > 0$ were obtained in a similar way. The results of both experiments performed with the same W1898 mixture in the same 3.1TN70 cells are presented in Figs. (**17** and **18**).

The transition voltages ($U_{TN} = 1.5$ V) from a twisted TN to an HT structure are the same. At the threshold U_{TN} value (see Figs. **17** and **18**), on the basis of eq. (1), one can calculate the factor $F_{TN} = \Delta\varepsilon\varepsilon_{o}U_{TN}^{2}$ and subsequently K_{TN} constant:

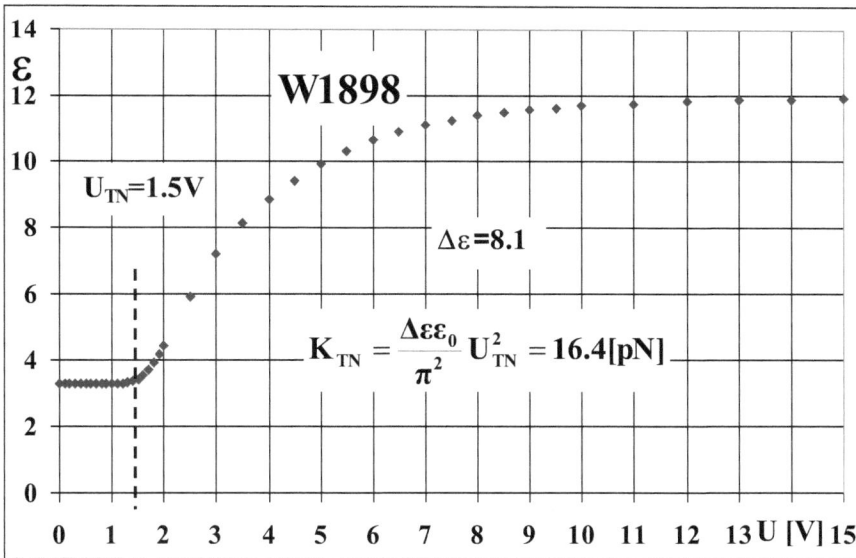

Fig. (**17**). Determination of the reduced constant K_{TN} for a TN effect in the W1898 mixture ($\Delta\varepsilon = 8.1$) by means of dielectric measurements ($f = 1$ kHz) in a 3.1TN70 cell.

$$K_{ii} = \pi^{-2}F_{ii} \tag{1}$$

One can see that K_{ii} as well as K_{TN} for MNCL mixtures with $\Delta\varepsilon > 0$ combined in Figs. (**9 - 18**) and Table **1** meet approximately the relation of:

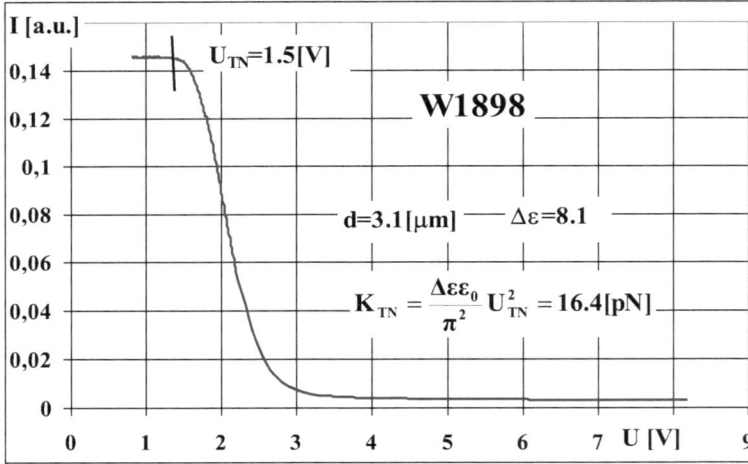

Fig. (18). Determination of the K_{TN} reduced constant for a TN effect by means of optical measurements ($\lambda = 0.589$ μm) in a 3.1TN70 cell filled with a W1898 mixture ($\Delta\varepsilon = 8.1$) placed between crossed polarizers. The polarization plane of incident light was parallel to director **n** at the input surface of a TN layer.

$$K_{TN} = K_{11} + \frac{1}{4}\left(K_{33} - 2K_{22}\right), \tag{2}$$

which relates K_{11}, K_{22} and K_{33} with K_{TN}. This confirms the consistency of methods used for the determination of elastic constants K_{ii} of tested MNCL mixtures with $\Delta\varepsilon > 0$.

3.5. Determination of Rotational Viscosity γ of W1898 Mixture

The rotational viscosity γ was estimated on the basis of measurements of the switching-on time $\tau_{on} \approx t_{0-90}$ of a TN electro-optical effect. Switching-on time τ_{on} is defined in this case as the time when light transmission of TN cell placed between two crossed polarizers, driven by voltage U pulse, changes from 0% to 90% (see Fig. **19**).

A TN cell with a cell gap d and the twist angle of $\xi = \pi/2$ was filled with a liquid crystalline mixture placed between crossed polarizers. The polarization plane of incident light was parallel to the direction of the director **n** on an incident side of the cell. The alternating voltage $U(t)$ of an amplitude U higher than "saturation voltage" of the TN effect was applied. In the case of W1898 mixture, a driving voltage pulse of rectangular shape with a period of 10 ms was applied.

A switching-on time τ_{on} was then assessed. Using the obtained switching-on time value $\tau_{on} \approx t_{0-90}$ one can calculate the rotational viscosity γ of the mixture by using Tadumi equation for a TN effect [11]:

$$\gamma = \frac{\tau_{ON}(\varepsilon_o \Delta\varepsilon U^2 - \pi^2 K_{TN})}{d^2}, \tag{3}$$

where: U is the amplitude of alternating driving voltage, K_{TN} is the reduced elastic constant for a TN effect with $\xi = \pi/2$. The result of the measurement of the rotational viscosity γ for the W1898 mixture at 25°C is shown in Fig. (**19**).

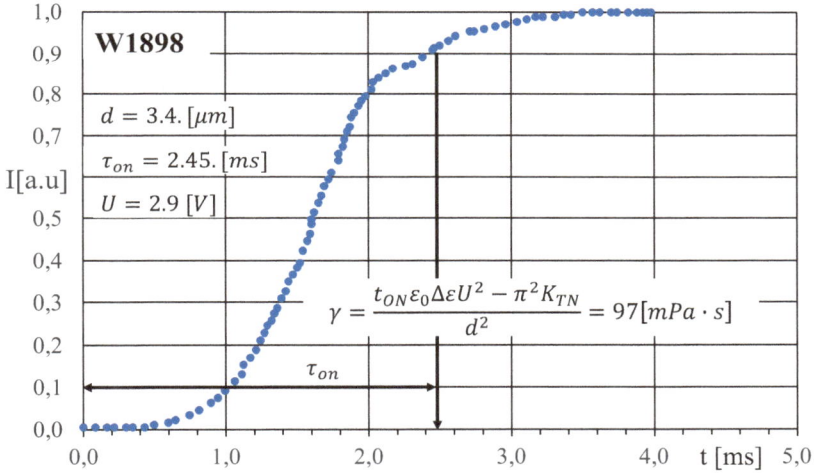

Fig. (19). An intensity I of a white light transmitted through a 3.4TN cell filled with W1898 mixture (at 25°C), placed between two crossed polarizers as a function of time t after applying driving voltage U pulse of a rectangular shape with a period of 10 ms. Polarization plane of incident light was parallel to the director $\boldsymbol{n}$.

The results of rotational viscosity γ study described above were verified by a study of switching-off time t_{off} during observation of an Electrically Controlled Birefringence (ECB) electro-optical effect in the HG cell filled with the same MNLC mixture. Switching-off time $t_{off} \approx t_{100-10}$ measured in an HG cell is connected with rotational viscosity γ by the following Tadumi equation [6] for an ECB effect:

$$\gamma = \frac{\tau_{OFF}\pi^2 K_{11}}{d^2} \tag{4}$$

The obtained values of rotational viscosities of seven MNCL mixtures, gathered in Table **1**, are the average results yield using Eqs. (3) and (4). It should be noted that the results obtained from the study of switching-on time t_{on} of a TN effect and switching-off time t_{off} of an ECB effect did not differ significantly (for W1898: $\gamma = 97$ mPa·s with TN, $\gamma = 76$ mPa·s with ECB) what proves the correctness of rotational viscosity γ determination in all MNLC mixtures under study. It is worth to point out that evaluated viscosities on the base of Tadumi equations are not strictly real viscosities of the given mixture. However, they are very useful to compare different working mixtures.

3.6. Studies of Physical Properties of Other MNLC Mixtures

The studies of physical properties presented above were made for all selected MNLC mixtures, so below we present results of dielectric and refractometric studies as well as a determination of Frank elastic constant K_{11} and reduced elastic constant K_{TN} summarized for above mixtures (Figs. **20-53**). The data presented below were obtained at the temperature of 25°C, except for temperature characteristics, whereas spectra T(λ) was registered mainly by means of a JASCO V670 spectrometer.

3.6.1. Study of W1820 Mixture

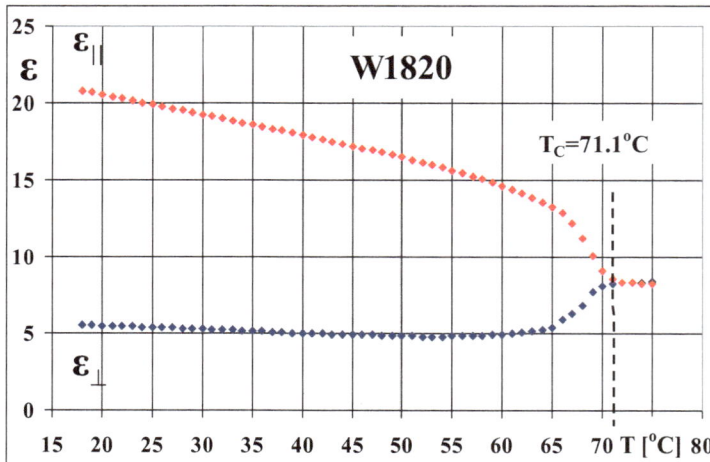

Fig. (20). Temperature characteristics of real parts $\varepsilon'_\parallel(T)$ and $\varepsilon'_\perp(T)$ for W1820 at $f = 1.0$ kHz determined using a 5.1HG10 and 5.0HTAu cells.

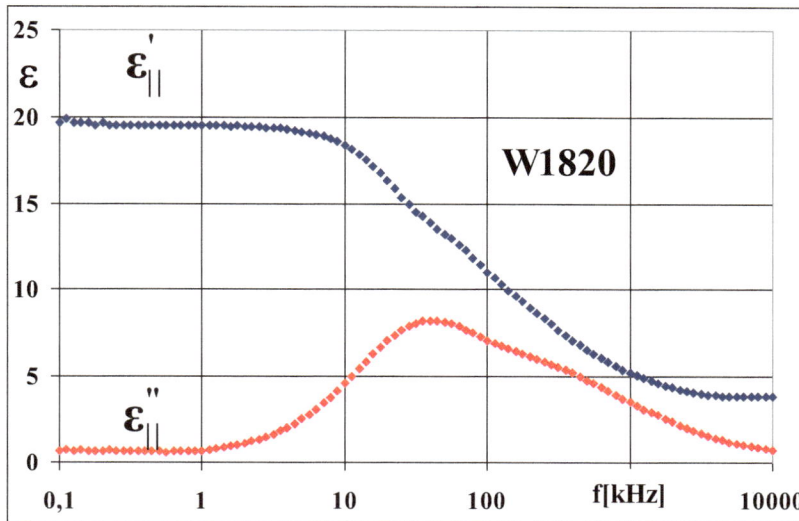

Fig. (21). Dispersion of real $\varepsilon'_\parallel(f)$ and imaginary $\varepsilon''_\parallel(f)$ parts of the permittivity measured for W1820 mixture in 5.0HTAu cell.

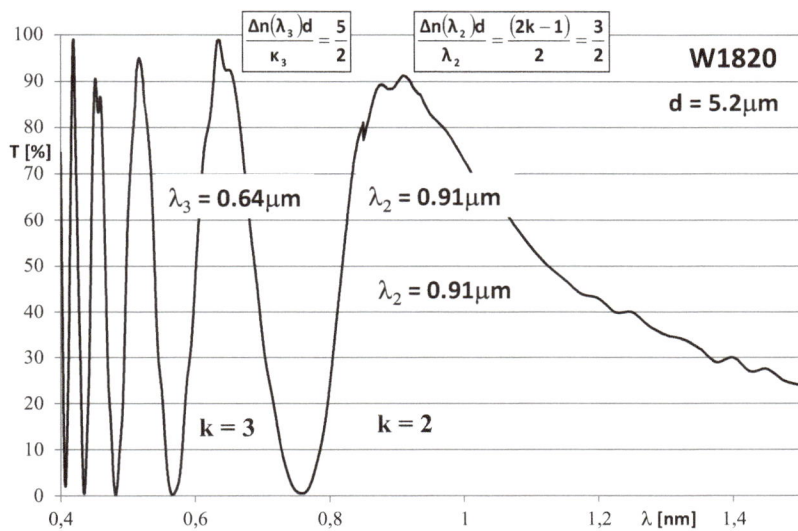

Fig. (22). Interference birefringence fringes in a 5.2HG500 cell filled with W1820 placed between crossed polarizers. This illustrates a $\Delta n(\lambda_k) = (2k-1)\ \lambda_k/(2d)$ relation [1, 2, 9, 10, 12, 13].

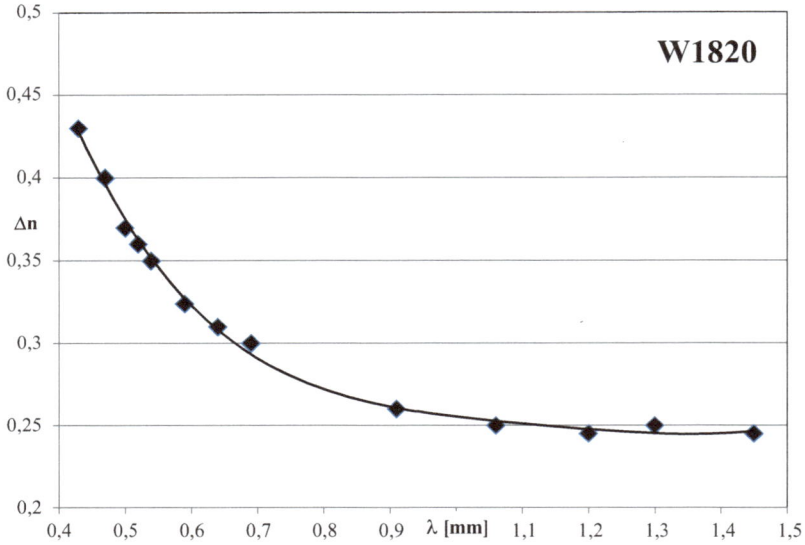

Fig. (23). The dispersion of the optical anisotropy $\Delta n(\lambda)$ for W1820. The graph illustrates results obtained by a combined method using the Abbe refractometer and interference in a wedge cell.

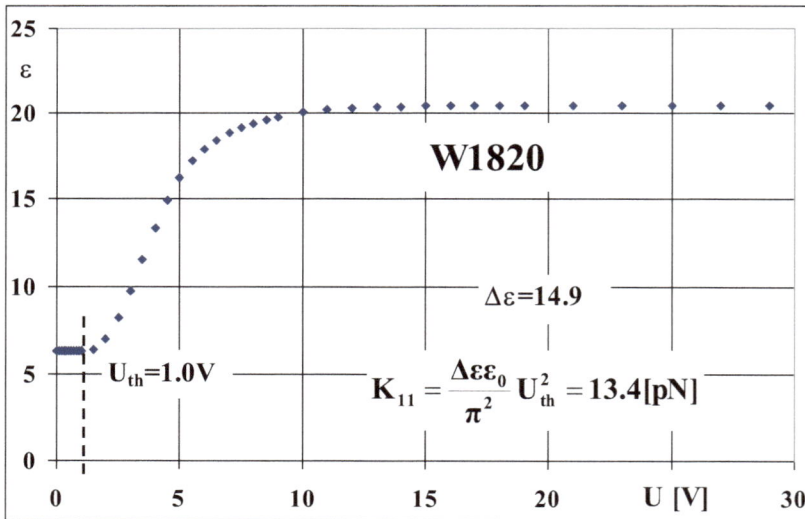

Fig. (24). Splay deformation of the W1820 monitored in a dielectric way ($f = 1$kHz) in a hybrid 5.1HIPSP40 cell. Voltage U was applied to electrodes 3 and 4 in section S of 5.1HIPSP40 cell (see Fig. **13**, Chapter II).

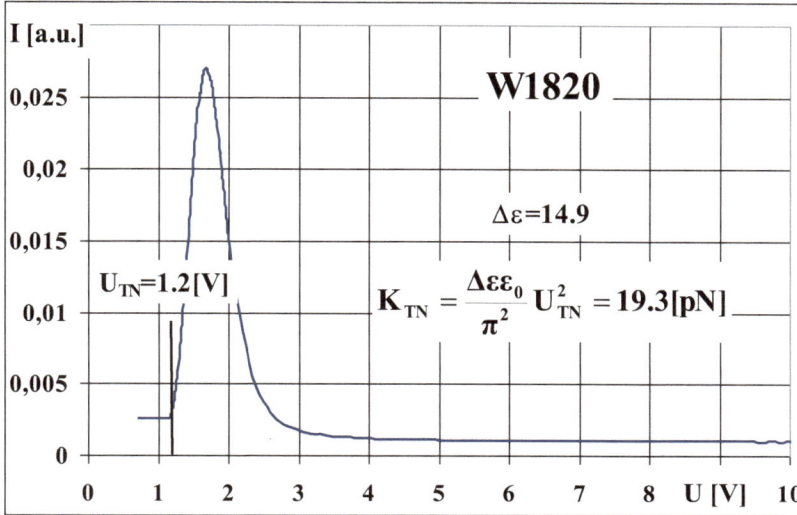

Fig. (25). Determination of the reduced elastic constant K_{TN} at a TN effect in the W1820 using optical measurements ($\lambda = 589\mu m$) in a 3.4TN70 cell. Polarization plane of an incident light formed the angle $\alpha = \pi/4$ with the director *n* at the incident side of the TN slab.

3.6.2. Study of W1852

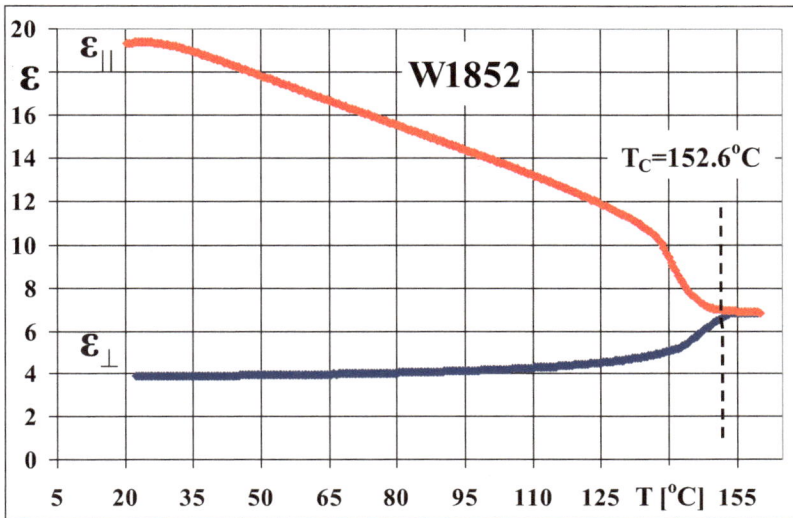

Fig. (26). Temperature characteristics of real parts of $\varepsilon'_{\parallel}(T)$ and $\varepsilon'_{\perp}(T)$ for the W1852 mixture at the frequency $f = 1.0$ kHz. $\varepsilon'_{\parallel}(T)$ and $\varepsilon'_{\perp}(T)$ were determined in 5.2HG10 and 5.2HTAu cells, respectively.

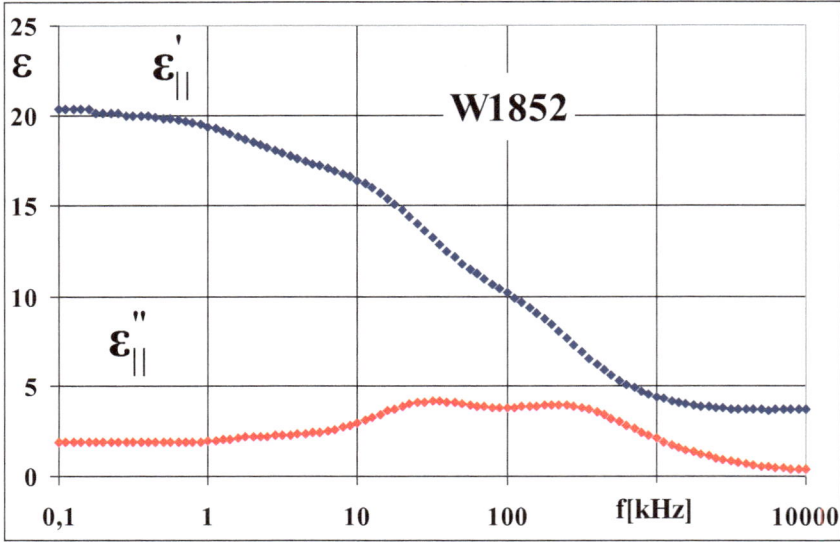

Fig. (27). The dispersion of real $\varepsilon_{\parallel}'(f)$ and imaginary parts $\varepsilon_{\parallel}''(f)$ of the W1852. These characteristics were determined using a 5.2HTAu cell.

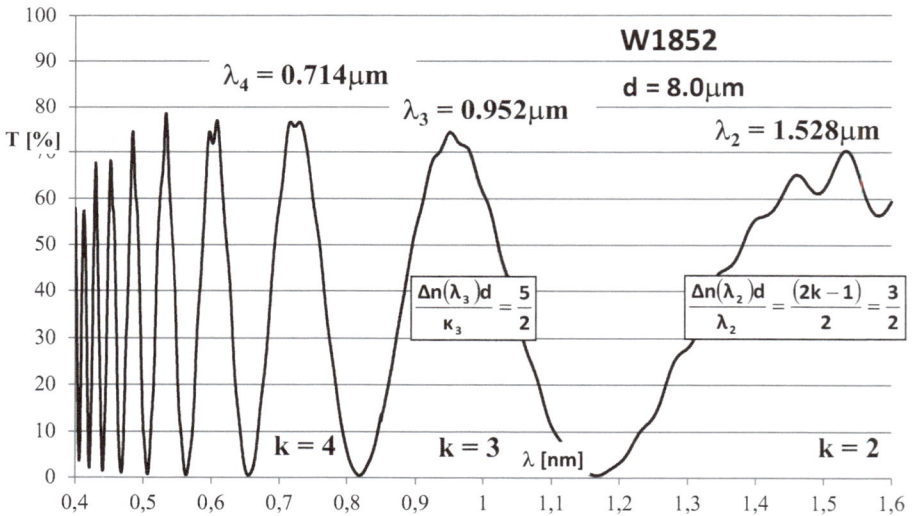

Fig. (28). Interference birefringence fringes observed in an 8.0HG500 cell filled with W1852 mixture. The cell was placed between crossed polarizers. This illustrates a relation $\Delta n(\lambda_k) = (2k-1)\, \lambda_k/(2d)$ [2, 12, 14, 15].

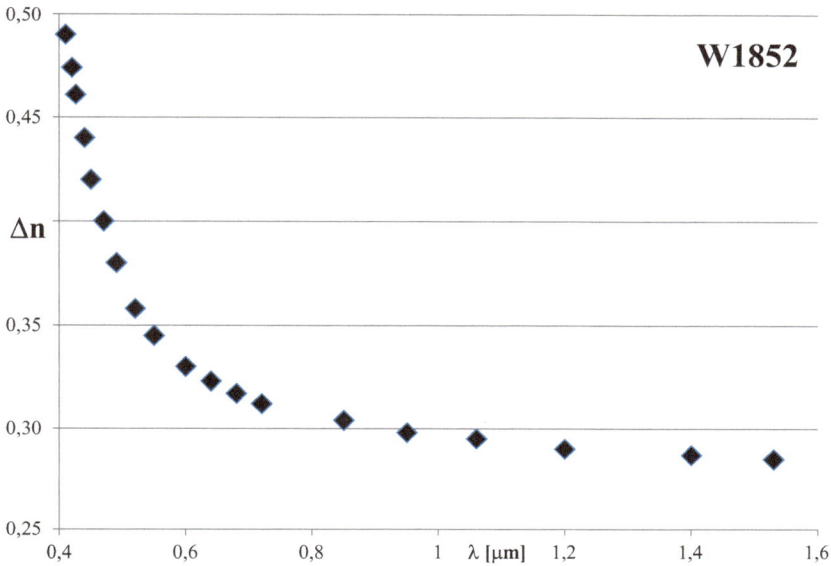

Fig. (29). The dispersion of the optical anisotropy $\Delta n(\lambda)$ of W1852 mixture. The graph illustrates results obtained by the combined method using Abbe refractometer and interference in the wedge cell.

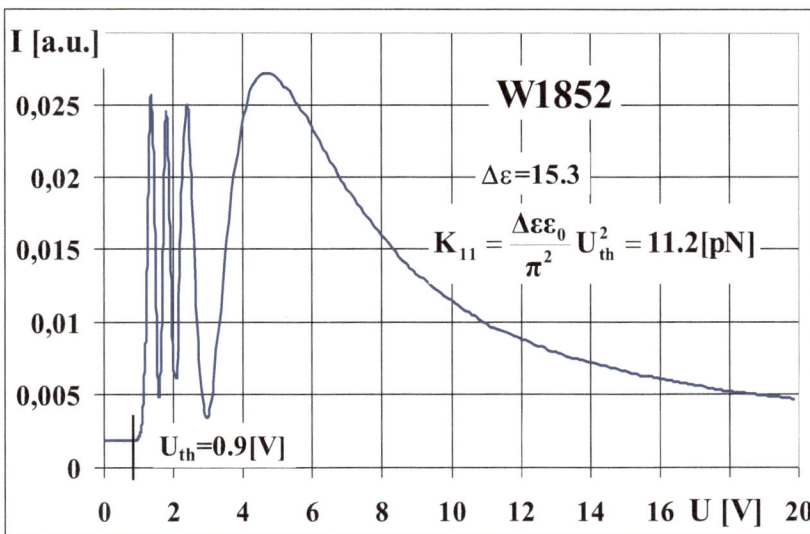

Fig. (30). Splay deformation of the W1852 mixture monitored in an optical way ($\lambda = 0.652\mu m$) in a hybrid 8.1HIPSP40 cell. Voltage U was applied to electrodes 3 and 4 in section S of an 8.1HIPSP40 cell (see Fig. **13**, Chapter II). The cell was placed between crossed polarizers. Polarization plane of incident light formed with the director **n** an angle $\alpha = \pi/4$.

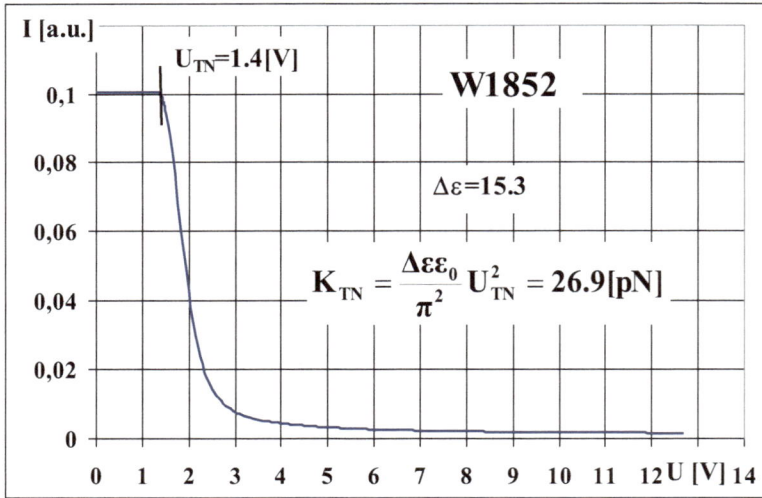

Fig. (31). Study of the reduced elastic constant K_{TN} of the W1852 mixture done at $\lambda = 0.656\ \mu m$ in a 3.0TN70 cell. Polarization plane of incident light was parallel to the director **n** at the incident side of TN slab.

3.6.3. Study of W1825 Mixture

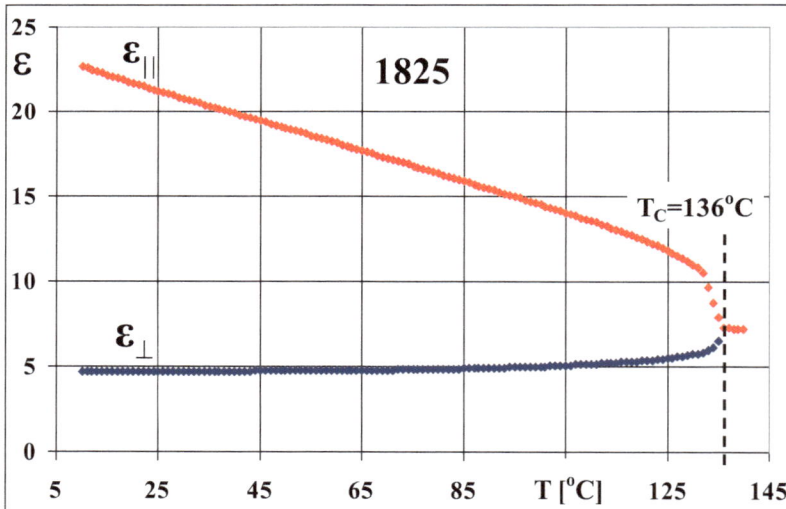

Fig. (32). Temperature dependence of $\varepsilon_\parallel'(T)$ and $\varepsilon_\perp'(T)$ real parts permittivity of the W1825 mixture at the frequency of $f = 1.0$ kHz determined in 5.1HG10 and 5.2HTAu cells, respectively.

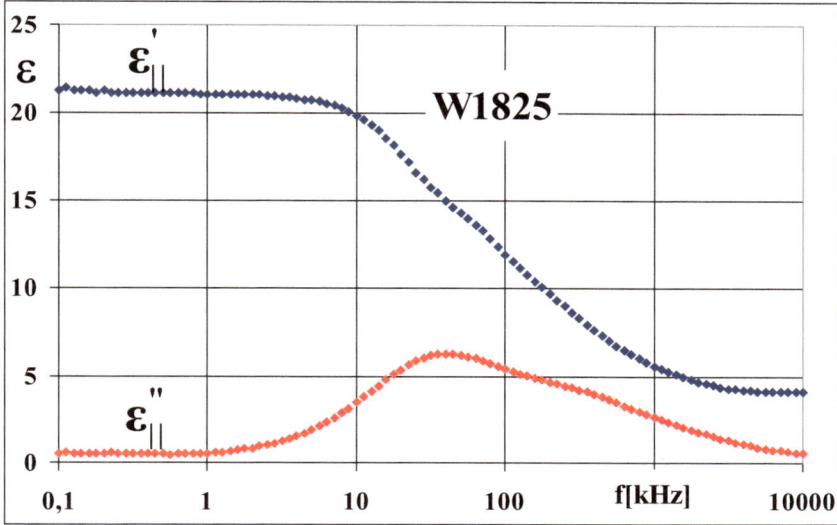

Fig. (33). The dispersion of real $\varepsilon'_\parallel(f)$ and imaginary $\varepsilon''_\parallel(f)$ parts of permittivity of W1825 mixture determined in a 5.2HTAu cell.

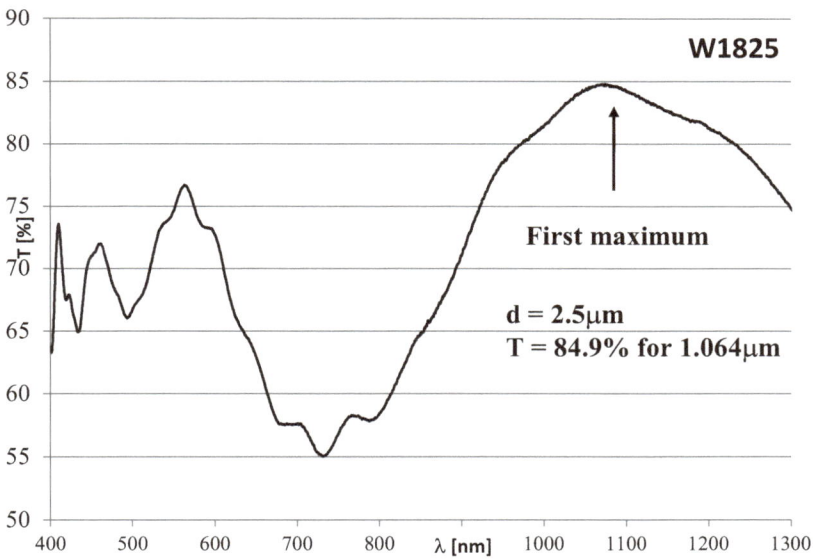

Fig. (34). Transmission T as a function of a wavelength λ for W1825 mixture in a 2.5TN500 cell placed between two crossed polarizers. Polarization plane of incident light was parallel to the direction of the director *n* at the incident side of a measuring cell.

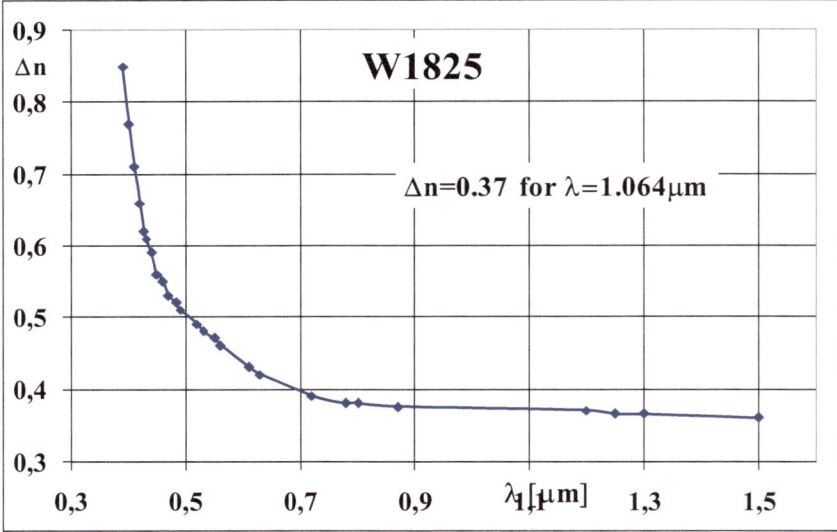

Fig. (35). The dispersion of the optical anisotropy $\Delta n(\lambda)$ of W1825 mixture. The graph illustrates the results obtained by the combined method using Abbe refractometer and interference in the wedge cell.

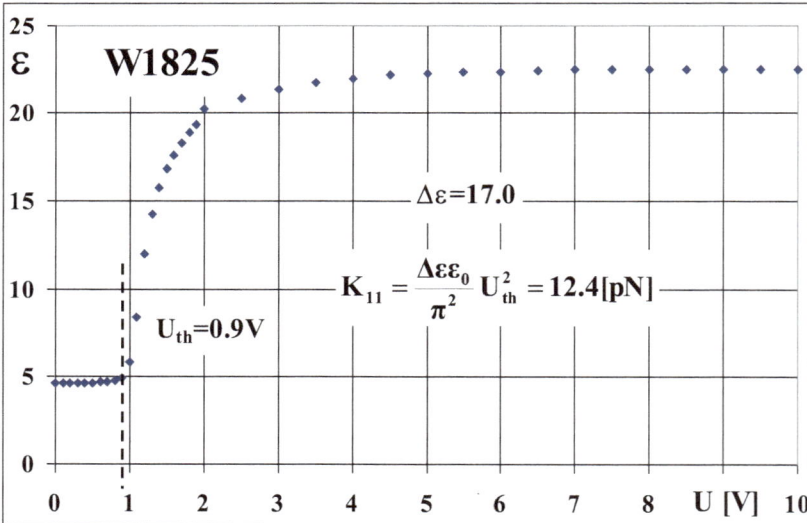

Fig. (36). Splay deformation in a structure of W1825 mixture monitored in an optical way ($f = 1$kHz) in a hybrid 5.1HIPSP20 cell. Voltage U was applied to electrodes 3 and 4 in section S of a 5.1HIPSP20 cell (see Fig. **13**, Chapter II).

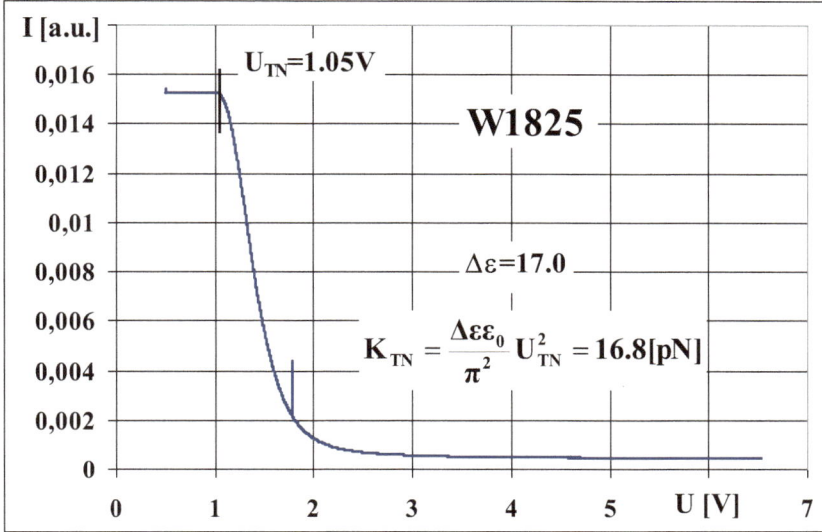

Fig. (37). Determination of the reduced elastic constant K_{TN} of W1825 mixture at $\lambda = 0.675$ μm in a 3.2TN70 cell; an observation of the threshold voltage U_{TN} at the TN electro-optical effect. Polarization plane of incident light was parallel to the director n at the incident side of the cell.

3.6.4. Study of W1791 Mixture

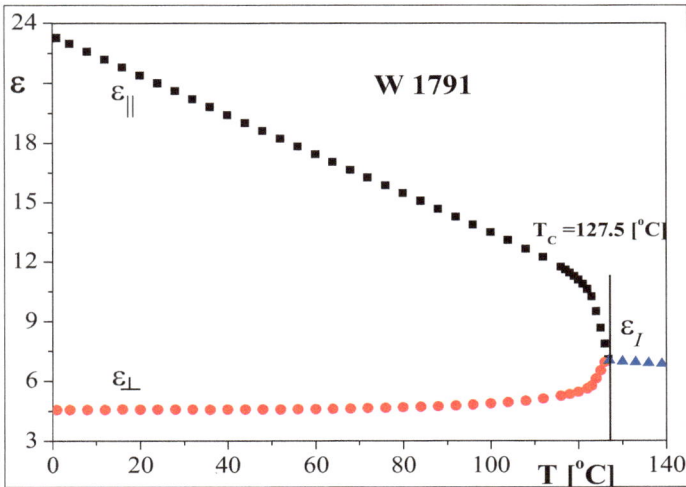

Fig. (38). Temperature dispersion of $\varepsilon'_{\parallel}(T)$ and $\varepsilon'_{\perp}(T)$ real parts of the permittivity of W1791 mixture at the frequency of $f = 1.0$ kHz determined in 5.3HGAu and 5.1HTAu cells, respectively.

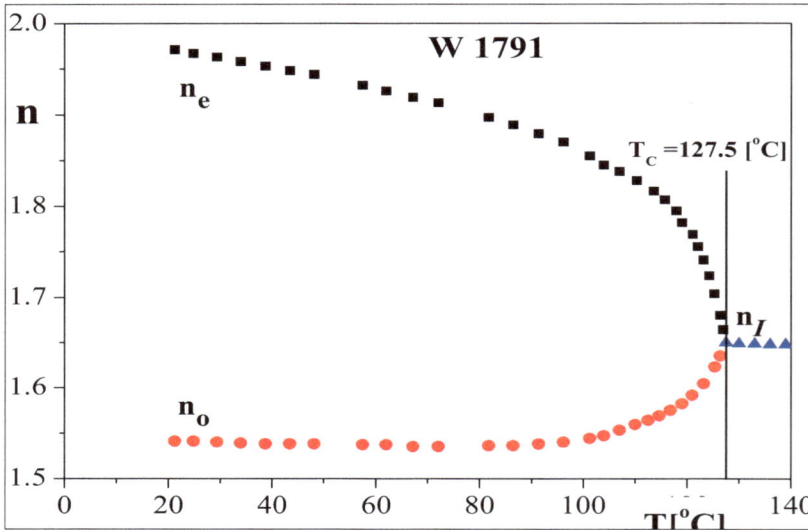

Fig. (39). Temperature characteristics of refractive indices $n_i(T)$, $n_o(T)$ and $n_e(T)$ of the W1791 mixture at $\lambda = 0.5893$ μm. The graph illustrates results obtained by the combined method using Abbe refractometer and observations of interference fringes in the wedge-shaped cell.

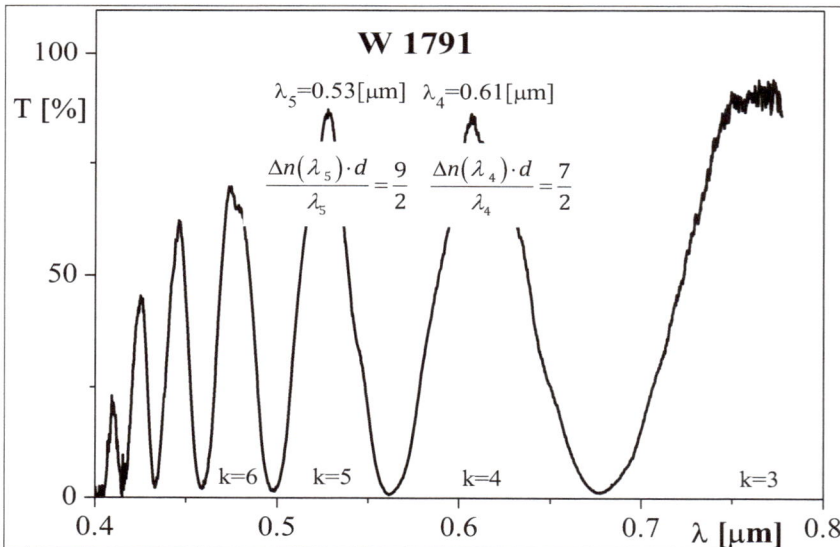

Fig. (40). An observation of interference birefringence fringes in a 5.1HG500 cell placed between crossed polarizers, filled with the W1791 mixture. The spectrum $T(\lambda)$ was registered by means of a BRC111A spectrometer. This illustrates a relation $\Delta n(\lambda_k) = (2k-1) \lambda_k/(2d)$ [10].

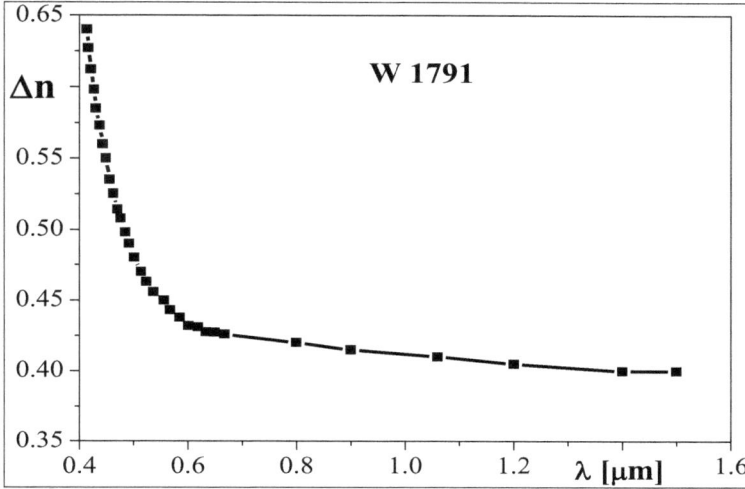

Fig. (41). Dispersion of the optical anisotropy $\Delta n(\lambda)$ of the W1791 mixture. The graph illustrates results obtained by the combined method using Abbe refractometer and interference in the wedge cell.

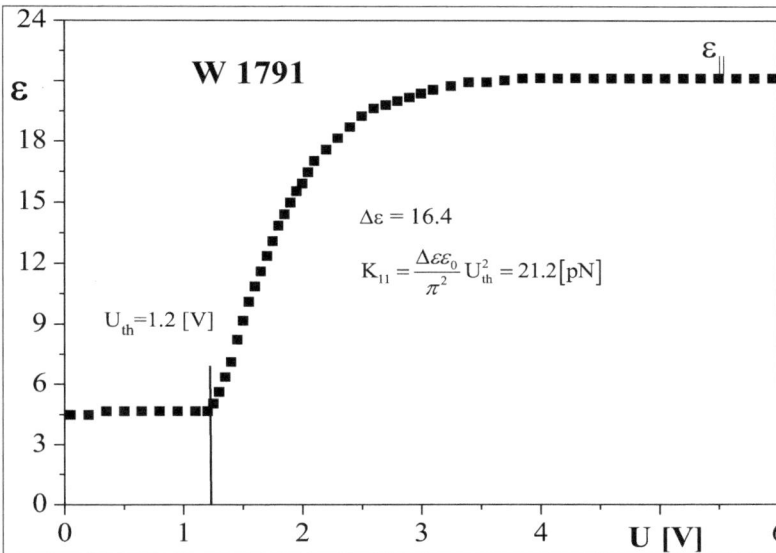

Fig. (42). The examination of the splay deformation in a structure of the W1791mixture monitored at the dielectric study ($f = 1$kHz) using a 5.3HIPSP40 hybrid cell. Voltage U was applied to electrodes 3 and 4 in section S of a 5.3HIPSP40 cell (see Fig. **13**, Chapter II).

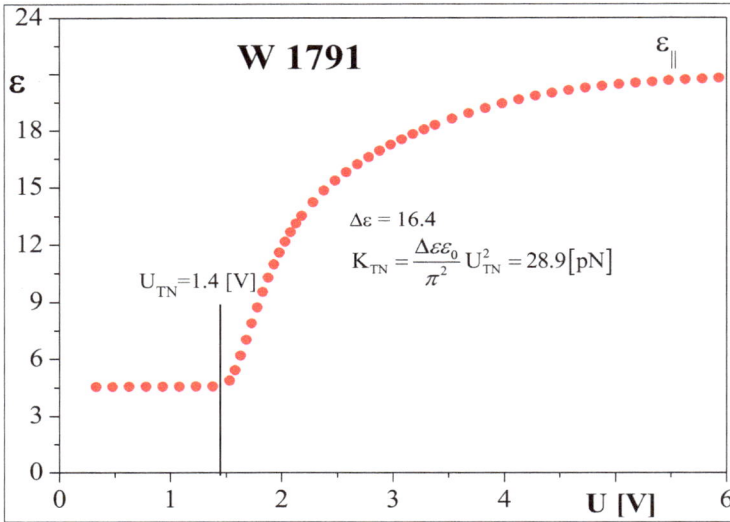

Fig. (43). Study of the reduced elastic constant K_{TN} of the W1791 mixture by means of dielectric measurements (f = 1kHz) in a 3.1TN70 cell.

3.6.5. Study of W1865 Mixture

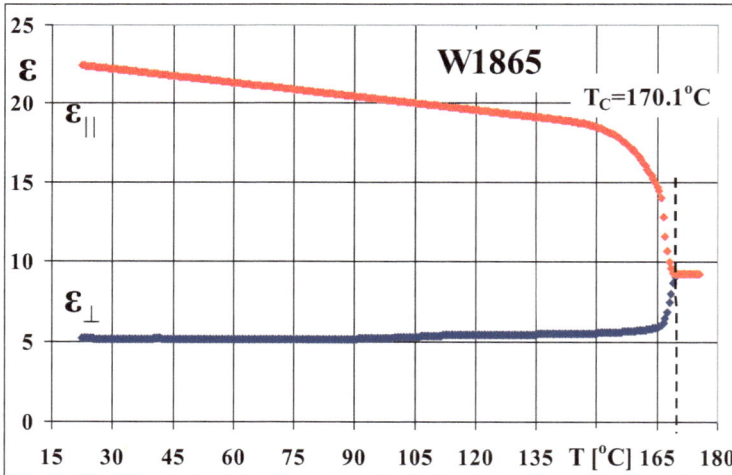

Fig. (44). Temperature characteristics of $\varepsilon'_{\parallel}(T)$ and $\varepsilon'_{\perp}(T)$ real parts of permittivity of the W1865 mixture at the frequency of f = 1.0 kHz. $\varepsilon'_{\parallel}(T)$ and $\varepsilon'_{\perp}(T)$ were determined in 5.3HG10 and 5.8HTAu cells, respectively.

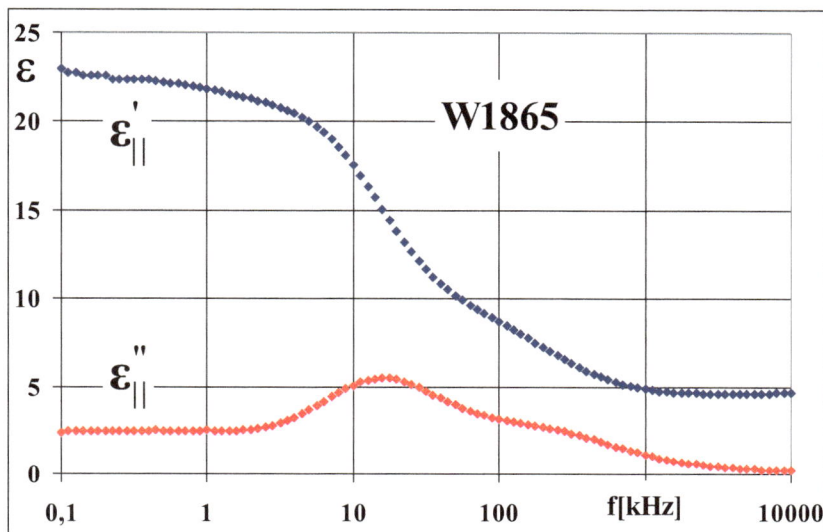

Fig. (45). The dispersion of real $\varepsilon_{\parallel}'(f)$ and imaginary $\varepsilon_{\parallel}''(f)$ parts of the permittivity of mixture W1865 determined using a 5.8HTAu cell.

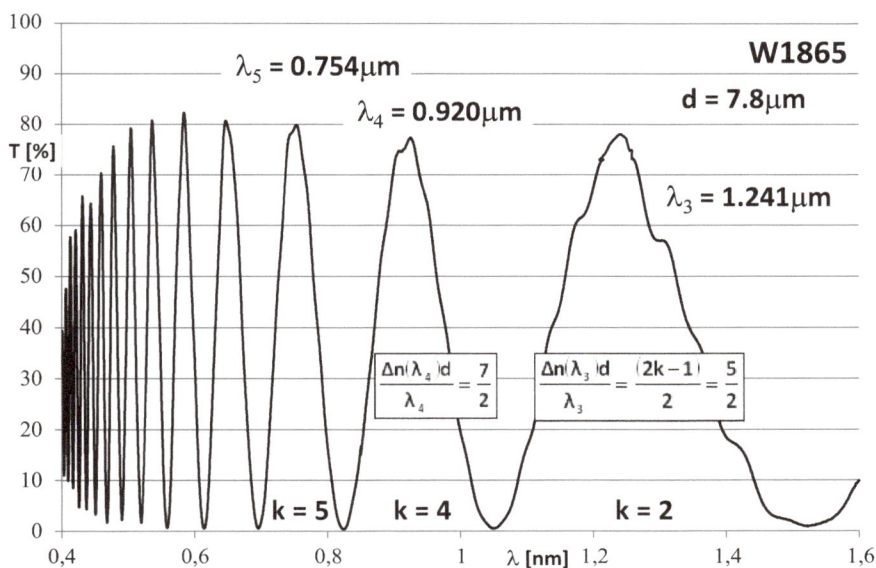

Fig. (46). An observation of interference birefringence fringes in a structure of the W1865 mixture in 7.8HG500 cell placed between crossed polarizers. This illustrates a relation $\Delta n(\lambda_k) = (2k-1)\,\lambda_k/(2d)$.

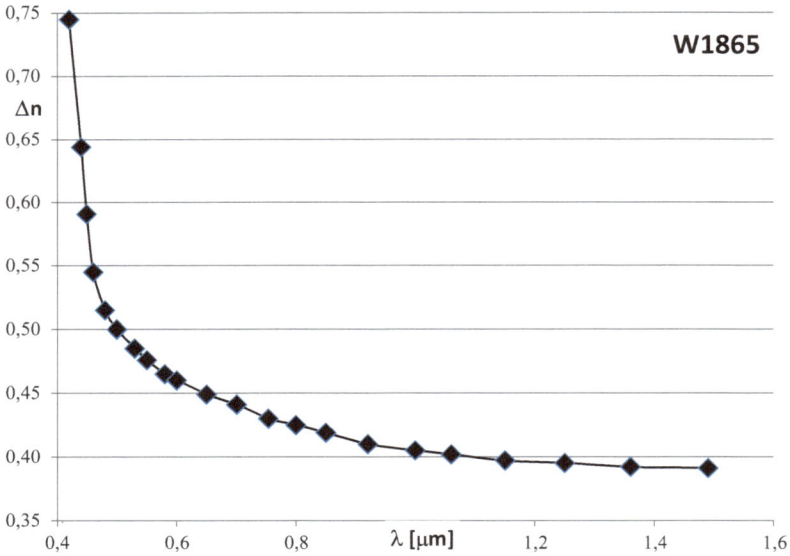

Fig. (47). The dispersion of the optical anisotropy $\Delta n(\lambda)$ of the W1865 mixture. The graph illustrates results obtained by the combined method using the Abbe refractometer and observation of interference fringes in the wedge-shaped cell.

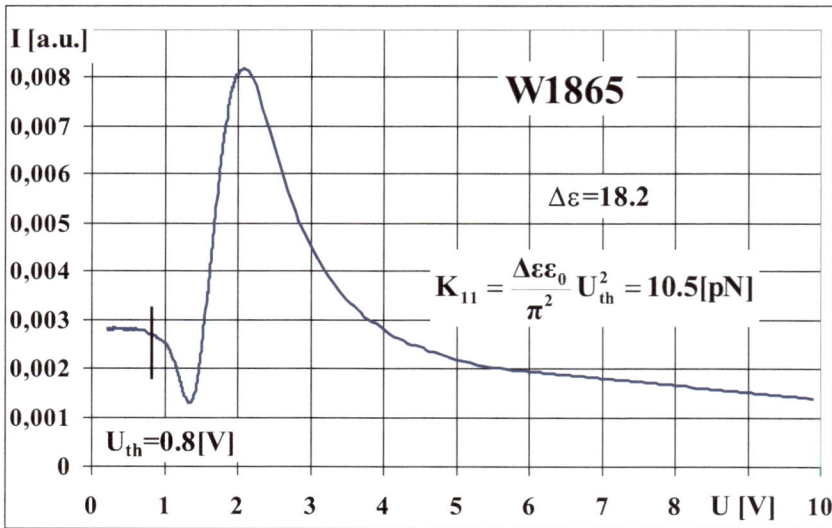

Fig. (48). A study of the splay deformation observed in the structure of the W1865 mixture at $\lambda = 0.652$ μm using a hybrid 1.9HG500 cell. The cell was placed between crossed polarizers. The polarization plane of incident light formed the angle of $\alpha = \pi/4$ with the director **n** of an HG slab.

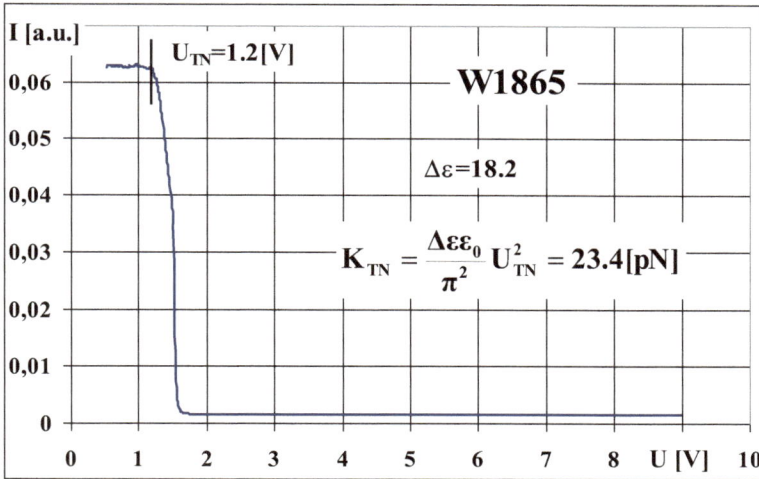

Fig. (49). A study of the reduced elastic constant K_{TN} of the W1865 mixture by means of optical measurements $\lambda = 0.675$ µm in a 2.3TN70 cell; an observation of the threshold voltage U_{TN} at the TN electro-optical effect. Polarization plane of incident light was parallel to the direction of the director n at the incident side of the measuring cell.

3.6.6. Study of W1115 Mixture

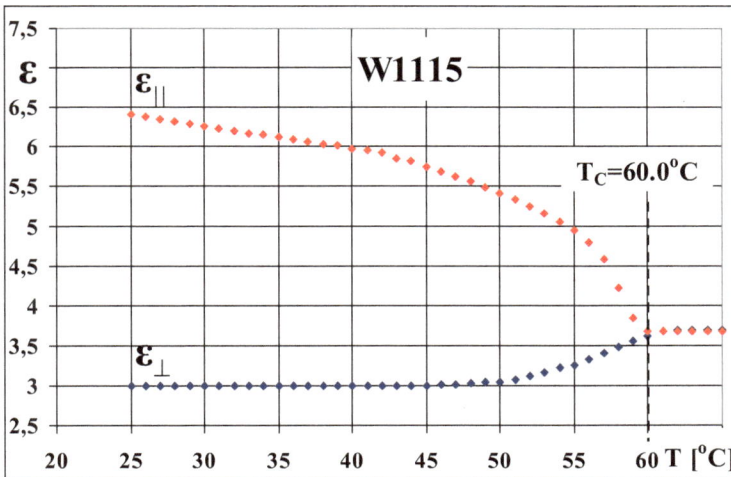

Fig. (50). Temperature characteristics of $\varepsilon'_\parallel(T)$ and $\varepsilon'_\perp(T)$ real parts of permittivity of the mixture W1115 observed at frequency $f = 1.0$ kHz. $\varepsilon'_\parallel(T)$ and $\varepsilon'_\perp(T)$ were specified in 6.0HG70 and 5.5HTAu, respectively.

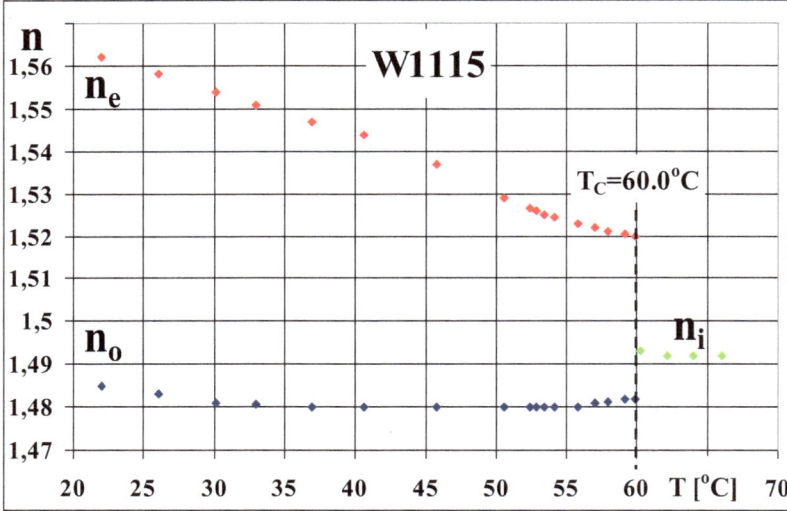

Fig. (51). The dispersion of refractive indices $n_i(T)$, $n_o(T)$ and $n_e(T)$ of the W1115 mixture at $\lambda = 0.5893$ µm. Points come from results obtained using the Abbe refractometer.

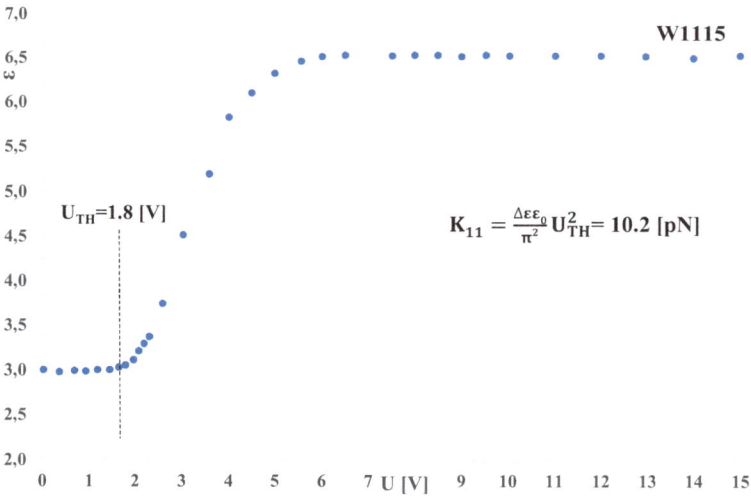

Fig. (52). Splay deformation of W1115 mixture monitored at the dielectric study (f = 1kHz) using a hybrid 6.0HIPSP40 cell. Voltage U was applied to electrodes 3 and 4 in section S of a 6.0HIPSP40 cell (see **Fig. 13**, Chapter II).

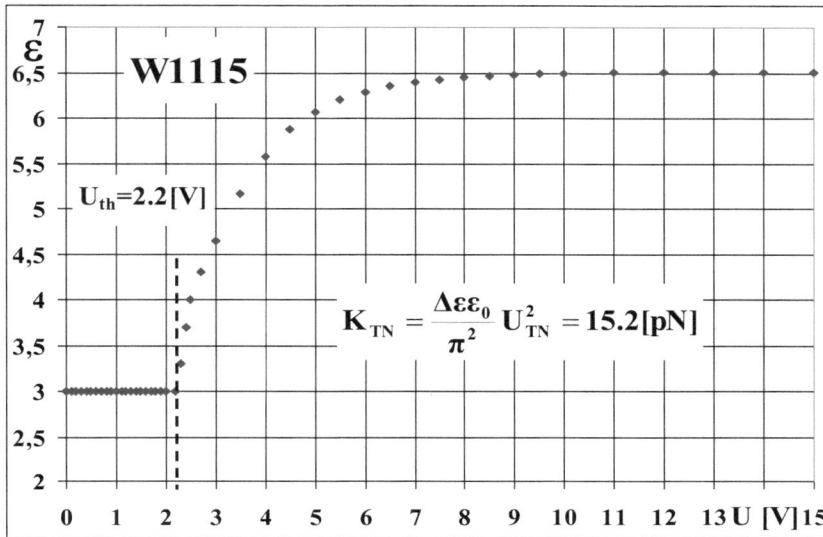

Fig. (53). Determination of the reduced K_{TN} elastic constant for W1115 mixture ($\Delta\varepsilon = 3.4$) at the dielectric study done at the frequency $f = 1$kHz. The observation of the threshold voltage U_{th} was done using a 6.1TN70 cell.

4. MATERIAL PARAMETERS OF MNLC MIXTURES

A series of studies described above provided detailed values of optical, dielectric and viscoelastic properties of eight examined MNLC mixtures. These results presenting evaluated values of such parameters like n_o, Δn, $\varepsilon_\perp$, $\Delta\varepsilon$, K_{11}, K_{22}, K_{33}, K_{TN} and γ are gathered in Table **1**. All MNLC mixtures presented in Table **1** were designed and composed for specific applications.

Table 1. Material parameters of MNLC mixtures observed at 25°C.

	W1795B	W1898	W1820	W1852	W1825	W1791	W1865	W1115
N-T[°C]-I	94.0	83.5	71.1	152.6	136.0	127.5	170.1	60.0
Cr-T[°C]-N	-5	-20	10	-10	-12	-20	10	-7
n_o ($\lambda = 0.589$ mm)	1.51	1.51	1.52	1.53	1.54	1.54	1.54	1.56
Δn ($\lambda = 0.589$ mm)	0.23	0.17	0.32	0.33	0.42	0.44	0.47	0.08
n_o ($\lambda = 1.064$ mm)	1.51	1.50	1.51	1.52	1.53	1.53	1.53	1.55
Δn ($\lambda = 1.064$ mm)	0.20	0.14	0.25	0.29	0.37	0.40	0.41	0.07
$\varepsilon_\perp$ ($f = 1.5$ kHz)	6.0	3.5	5.5	4.1	4.7	4.5	5.0	3.0

	W1795B	W1898	W1820	W1852	W1825	W1791	W1865	W1115
$\varepsilon_{\parallel}$ (f=1.5 kHz)	-2.5	8.1	19.5	15.3	17.0	16.4	18.2	3.5
K_{11} [pN]	15.9	12.0	13.5	11.2	12.5	21.2	10.5	10.2
K_{22} [pN]	7.9	7.5	8.2	7.7	7.4	8.3	7.8	7.5
K_{33} [pN]	18.2	24.1	33.0	31.0	32.1	25.2	29.4	30.0
K_{TN} [pN]	-	16.4	19.7	26.9	16.8	28.9	23.4	15.2
Γ[mPa·s]	23	10	18	-	31	31	-	40
γ[mPa·s]	150	97	180	320	284	155	595	165

Most of them were tested as functional materials used for manufacturing of specialized light modulators. These advanced, working modulators were designed, fabricated, and tested for extended spectral range and at a broad temperature range. Tests done, among the others, yield the following conclusions:

1. Due to high optical birefringence ($\Delta n > 0.3$ at VIS-NIR spectral ranges) and permittivity anisotropies ($\Delta \varepsilon$) accompanied with pronounced values of elastic constants K_{ii} and K_{TN} as well as due to the low viscosity coefficient γ the mixtures W1820, W1852, W1825, W1791 and W1865 can be applied for manufacturing of fast operating light modulators working at VIS-NIR spectral range. For example:
   a. by using the NLC mixture W1825 with $\Delta n = 0.37$ (at $\lambda = 1.064$ μm) precisely tuned to operate at the first interference maximum of a positive TN mode ($d = 2.5$ μm) a fast operating Liquid Crystal Cell (LCC) for laser rangefinder used for land, maritime, air or space navigation purposes (see Chapter 4) was manufactured;
   b. by using the NLC mixture W1820 with $\Delta n = 0.28$ (at $\lambda = 0.808$ μm) precisely tuned to operate at the first interference maximum of a positive TN mode ($d = 2.5$ μm), a fast liquid crystal switch, abbreviated 0.81LCN, for land, maritime, air or space navigation or communication purposes (including GPS systems) was manufactured;
   c. by using the NLC mixture W1791 with high optical anisotropy $\Delta n > 0.40$ a fast liquid crystal spectral filter (LCS filter, based on the ECB effect) selecting wavelength λ within VIS range was fabricated (see Chapter 6).
2. Due to the tunability of the optical anisotropy $\Delta n \sim 0.16$ in the center of the VIS range, and owing to relatively high the dielectric anisotropy $\Delta \varepsilon \sim 8$ and to pronounced values of elastic constants K_{ii} and K_{TN} accompanied with the low viscosity coefficient γ the NLC W1898 mixture can be applied in a new generation, bright glasses for the stereoscopic vision working at a TN mode [16].

3. Due to the tunability of the optical anisotropy $\Delta n \sim 0.08$ at the middle of VIS range (at $\lambda = 0.555$ µm) and owing to a relatively low value of an average permittivity ($\varepsilon_{av} = \varepsilon_{\perp} + \Delta\varepsilon/3 = 4.5$) accompanied with pronounced values of elastic constants K_{ii} and K_{TN} and very low viscosity coefficient γ the W1115 mixture can be applied for manufacturing a light valve working at TN mode; such a valve can be applied for the eye protection devices (*i.e.* Automatic Welding Helmet) characterized with low electric power consumption (see Chapter 6).

4. Due to the tunability of the optical anisotropy $\Delta n \sim 0.20$ at the middle of VIS range and owing to a high negative value of permittivity anisotropy $\Delta\varepsilon \sim -2.5$, accompanied with the pronounced values of elastic constants K_{ii} and K_{TN} and very low viscosity coefficient γ the W179 mixture can be applied in very fast liquid crystal light modulators working at a VAN mode (Vertical Aligned Nematic) at $d \sim 1.5$ µm.

CONSENT FOR PUBLICATION

Not applicable.

CONFLICT OF INTEREST

The author(s) confirms that there is no conflict of interest.

ACKNOWLEDGEMENTS

Declared none.

REFERENCES

[1]　E. Nowinowski-Kruszelnicki, J. Kędzierski, Z. Raszewski, L. Jaroszewicz, R. Dąbrowski, W. Piecek, P. Perkowski, M. Olifierczuk, K. Garbat, M. Sutkowski, E. Miszczyk, K. Ogrodnik, P. Morawiak, M. Laska, and R. Mazur, "High birefringence liquid crystal mixtures for lc electro-optical devices", *Opt. Appl.*, vol. 42, no. 1, pp. 167-180, 2012.

[2]　R. Dąbrowski, "New liquid crystalline materials for photonic applications", *Mol. Cryst. Liq. Cryst. (Phila. Pa.)*, vol. 421, pp. 1-21, 2004.
[http://dx.doi.org/10.1080/15421400490501112]

[3]　R. Dąbrowski, J. Dziaduszek, A. Ziółek, Ł. Szczuciński, Z. Stolarz, G. Sasnouski, V. Bezborodov, W. Lapanik, S. Gauza, and S-T. Wu, "Low viscosity, high birefringence liquid crystalline compounds and mixtures", *Opto-Electron. Rev.*, vol. 15, pp. 47-51, 2007.
[http://dx.doi.org/10.2478/s11772-006-0055-4]

[4]　P. Perkowski, D. Łada, K. Ogrodnik, J. Rutkowska, W. Piecek, and Z. Raszewski, "Technical aspects of dielectric spectroscopy measurements of liquid crystals", *Opto-Electron. Rev.*, vol. 16, no. 3, pp. 271-276, 2008.
[http://dx.doi.org/10.2478/s11772-008-0008-1]

[5]　P. Perkowski, "Dielectric spectroscopy of liquid crystals. Theoretical model of ITO electrodes influence on dielectric measurements", *Opto-Electron. Rev.*, vol. 17, no. 2, p. 180, 2009.
[http://dx.doi.org/10.2478/s11772-008-0062-8]

[6]   Z. Raszewski, "Measurement of permittivity of liquid-crystalline substances", *Electron. Technol.,* vol. 20, p. 99, 1987.

[7]   J.W. Baran, Z. Raszewski, J. Kędzierski, and J. Rutkowska, "Some physical properties of mesogenic 4-(trans-4'-n-alkylcyclohexyl)isothiocyanatobenzenes", *Mol. Cryst. Liq. Cryst. (Phila. Pa.),* vol. 123, pp. 237-245, 1985.
[http://dx.doi.org/10.1080/00268948508074781]

[8]   F. Zhang, L. Kang, Q. Zhao, J. Zhou, X. Zhao, and D. Lippens, "Magnetically tunable left handed metamaterials by liquid crystal orientation", *Opt. Express,* vol. 17, no. 6, pp. 4360-4366, 2009.
[http://dx.doi.org/10.1364/OE.17.004360] [PMID: 19293863]

[9]   J. Kędzierski, Z. Raszewski, M.A. Kojdecki, E. Nowinowski-Kruszelnicki, P. Perkowski, W. Piecek, E. Miszczyk, J. Zieliński, P. Morawiak, and K. Ogrodnik, "Determination of ordinary and extraordinary refractive indices of nematic liquid crystal by using wedge cells", *Opto-Electron. Rev.,* vol. 18, no. 2, pp. 214-218, 2010.
[http://dx.doi.org/10.2478/s11772-010-0009-8]

[10]   E. Miszczyk, Z. Raszewski, J. Kędzierski, E. Nowinowski-Kruszelnicki, M.A. Kojdecki, P. Perkowski, W. Piecek, and M. Olifierczuk, "Interference method for determination of refractive indices of liquid crystal", *Mol. Cryst. Liq. Cryst. (Phila. Pa.),* vol. 544, pp. 22-36, 2011.
[http://dx.doi.org/10.1080/15421406.2011.569262]

[11]   K. Tarumi, U. Frinkenzeller, and B. Schuler, "Dynamic behaviour of twisted nematic liquid crystals", *Jpn. J. Appl. Phys.,* vol. 31, pp. 2829-2836, 1992.
[http://dx.doi.org/10.1143/JJAP.31.2829]

[12]   J. Kędzierski, Z. Raszewski, E. Nowinowski-Kruszelnicki, M.A. Kojdecki, W. Piecek, P. Perkowski, and E. Miszczyk, "Composite method for measurement of splay and bend nematic constants by use of single special in-plane switched cell", *Mol. Cryst. Liq. Cryst. (Phila. Pa.),* vol. 544, pp. 57-68, 2011.
[http://dx.doi.org/10.1080/15421406.2011.569273]

[13]   E. Nowinowski-Kruszelnicki, J. Kędzierski, Z. Raszewski, L. Jaroszewicz, R. Dąbrowski, M.A. Kojdecki, W. Piecek, P. Perkowski, M. Olifierczuk, E. Miszczyk, K. Ogrodnik, P. Morawiak, and K. Garbat, "Measurement of elastic constants of nematic liquid crystal with use of hybrid in-plan--switched cell", *Opto-Electron. Rev.,* vol. 20, no. 3, pp. 255-259, 2012.
[http://dx.doi.org/10.2478/s11772-012-0027-9]

[14]   E. Nowinowski-Kruszelnicki, E. Walczak, A. Kieżun, and L.R. Jaroszewicz, "Light transmission loss in liquid crystal waveguides", *Proc. SPIE,* vol. 3318, pp. 410-413, 1998.
[http://dx.doi.org/10.1117/12.300014]

[15]   J. Kędzierski, M.A. Kojdecki, Z. Raszewski, J. Zieliński, and L. Lipińska, "Determination of anchoring energy, diamagnetic susceptibility anisotropy, and elasticity of some nematics by means of semiempirical method of self-consistent director field", *Proc. SPIE,* vol. 6023, pp. 26-40, 2005.
[http://dx.doi.org/10.1117/12.648167]

[16]   R. Mazur, W. Piecek, Z. Raszewski, P. Morawiak, K. Garbat, O. Chojnowska, M. Mrukiewicz, M. Olifierczuk, J. Kedzierski, and R. Dabrowski, "Nematic liquid crystal mixtures for 3D active glasses application", *Liq. Cryst.,* vol. 44, pp. 417-426, 2017.

CHAPTER 4

Liquid Crystal Cell for Space-borne Laser Rangefinder

Leszek Roman Jaroszewicz and **Zbigniew Raszewski**[*]

Military University of Technology, Warsaw, Poland

Abstract: Liquid Crystal Cell (LCC) for space-borne laser rangefinder for space mission applications was developed, manufactured and tested at the Military University of Technology (MUT)Warszawa, Poland, in cooperation with Vavilov State Optical Institute (Vavilov SOI), Petersburg, Russia. LCC operated at the positive TN mode (d=2.5 μm) tuned for the laser beam of a wavelength of λ=1064 nm. It switched the polarization plane of the laser beam at the beam's energy density not smaller than 0.15 J/cm^2 at the pulse duration about 8 ns. At the working aperture, not less than 15 mm the transmission T of LCC was not smaller than 95%. Switching-on and switching-off times were assessed for LCC driven with a voltage of an amplitude U=10 V. At the temperature range from 20°C to 40°C, measured switching-on and switching-off times were not larger than 0.7 ms and 7 ms, respectively. The LLCs developed at the MUT were tested at the Vavilov SOI under procedure dedicated to space equipment. LLCs were mounted in the laser rangefinder of the space lander serving while the "*Phobos-Grunt*" mission was launched on November 8th, 2011, in Kazakhstan.

Keywords: And twisted alignment, Contrast ratio, Dielectric anisotropy, Homogeneous, Liquid crystal cell, Optical anisotropy, Ordinary refractive index, Rotational viscosity, Twisted Nematic effect.

1. INTRODUCTION

Liquid crystal cell (LCC) for space-borne laser rangefinder was developed, manufactured, and tested at the Military University of Technology (MUT), Warsaw, Poland [1]. It was scrupulously tested in the laboratories of Vavilov State Optical Institute (VSOI) in Sankt Petersburg, Russian Federation, as well as a laser rangefinder during field tests. Two switches of polarization state; LCC1 and LCC2 (Fig. **1**), form the part of a system generating four (1, 2, 3 and 4), periodically sent, laser light beams in a rangefinder module of the International

[*] **Corresponding author Zbigniew Raszewski:** Military University of Technology, Faculty of Advanced Technologies and Chemistry, Warsaw, Poland; E-mail: zbigniew.raszewski@wat.edu.pl

Space Mission "Phobos-Grunt" of Space Agency of the Russian Federation. The mission was to place a return module on the Mars' moon – Phobos, then to collect samples of soil and bring them to Earth.

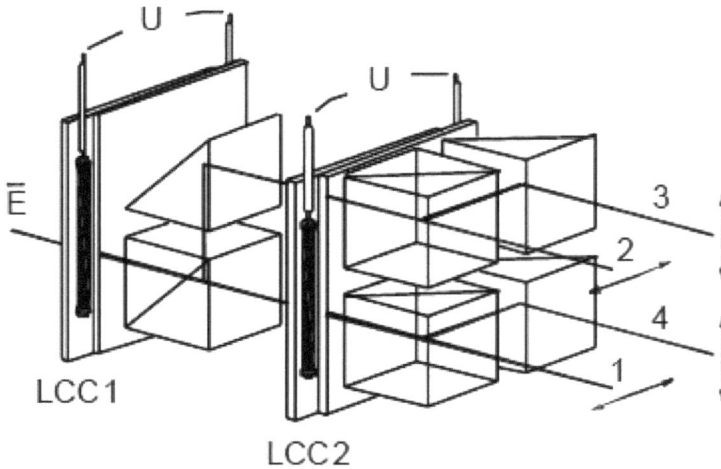

Fig. (1). LCC1 and LCC2 in rangefinder's optical system generating four (1, 2, 3 and 4) alternating laser beams with *E* vector marked with arrows.

The rangefinder of the space landing module sends pulses of a beam 1 in the direction to the landing place. The other three beams are sent along three edges of an equilateral pyramid of which height is the beam 1. Laser pulses run from the landing module to landing place and then return to the rangefinder. By measuring lasers pulsed times of flight, one obtains information on both; rangefinder distance from its landing place and its orientation in space. The probe with two LCCs designed and manufactured at MUT, Warsaw, Poland as a result of a Key Project POIG.01.03.01-14-016/08 "*New Photonic Materials and their Advanced Application*" was launched on November 8th, 2011, in Kazakhstan.

2. LCC TECHNICAL REQUIREMENTS

According to technical requirements, LCC should:

- switch the polarization plane of the laser beam, at $\lambda = 1064$ nm, by an angle of $\xi = \pi/2$,
- transfer energy density not lower than 0.15 J/cm^2 of laser radiation with a pulse duration of 8 ns with a repetition rate of 100 Hz,
- withstand at least 10^6 pulses,

- have the aperture Φ not lower than 15 mm,
- have the transmission T not lower than 95%,
- have the switching-on time τ_{on} not longer than 1.5 ms (1.0 ms expected),
- have the switching-off time τ_{off} not longer than 10.0 ms (expected 7.5 ms),
- operate at the driving voltage $U_{RMS} < 18$ V at the frequency $f = 1500$ Hz,
- operate properly at the temperature range from 20°C to 40°C,
- operate in the vacuum (10^{-7} Tr),
- exhibit a deformation of a wavefront below $\lambda/2$,
- withstand g-force up to 40g,
- be resistant to "cosmic radiation" to a total dose not less than 10^5 rads,
- endure storage temperature from -10°C to +50°C and operate properly during a four-year mission.

It was decided to develop an LCC based on the TN (Twisted Nematic) effect, which at proper design and manufacturing ensures the rotation of the polarization plane of an incident beam by an angle of $\xi = \pi/2$ exactly.

3. ELECTRO-OPTICAL TWISTED NEMATIC EFFECT

3.1. Configuration and Transmission of a Liquid Crystal TN Transducer

One creates a twisted nematic structure (TN) if the molecular director $\boldsymbol{n}$ in the NLC slab of thickness d confined between two bounding surfaces is twisted by an angle of $\xi = \pi/2$. The bounding surfaces of the cell inducing TN usually are made of glass substrates equipped with transparent Indium Tin Oxide (ITO) electrodes coated with rubbed aligning layers. The orientations of the director $\boldsymbol{n}$ ensured with rubbed aligning layers make the angle $\xi = \pi/2$ at two opposite boundary surfaces. NLC or MNLC mixture design for operation at the TN regime should exhibit the positive optical ($\Delta n = n_e - n_o > 0$) as well as the dielectric ($\Delta \varepsilon = \varepsilon_{\parallel} - \varepsilon_{\perp} > 0$) anisotropies.

Let us consider a polarized light beam of the wavelength λ at normal incidence onto the boundary surface of TN structure. The polarization plane of such a beam can be regarded as perpendicular or parallel to the director $\boldsymbol{n}$ at the incident side of the TN slab. The polarization plane of a beam of wavelength λ passing through the TN slab of particular thickness d and optical anisotropy Δn rotates by the angle $\xi = \pi/2$ (Eq. 2). The driving voltage U applied to the transparent ITO electrodes induces an electric field $\boldsymbol{E}$ in the space between them hence affects the NLC with $\Delta \varepsilon > 0$. An induced force momentum $F \sim \Delta \varepsilon \, E^2$ drives the director $\boldsymbol{n}$ orientation within the cell from twisted (TN) to homeotropic (HT) one. Since, at the HT director structure, the long axis of the optical indicatrix (n_e) of the medium is parallel to the light path, the beam experiences an optically isotropic medium.

Hence, the light polarization plane remains unchanged. One can conclude that at the voltage enforced HT structure, the effective optical anisotropy of the medium is suppressed ($\Delta n_{ef} = 0$).

Let us consider TN slab placed between a two crossed, "ideal" polarizers - called here as polarizer (P) and analyzer (A). Let us assume that an axis of polarizer P is parallel to the director n at the incident side of the TN structure and the transmission axis of the analyzer A is parallel to the director n at the exit surface. Hence, the coefficient of transmission T is given by the formula [2 - 4]:

$$T = 1 - \frac{sin^2 \left[\frac{\pi}{2} \sqrt{1 + (\frac{2d\Delta n}{\lambda})^2} \right]}{1 + (\frac{2d\Delta n}{\lambda})^2} \tag{1}$$

The maximum of the transmission T of such a passive TN mode described above appears when:

$$\frac{2d\Delta n}{\lambda} = \sqrt{4k^2 - 1} \tag{2}$$

where an integer number k = 1, 2, 3... describes an order of the interference maximum.

Fig. (**2**) presents the dispersion of the transmission T of a TN passive mode for two values of the factor $d\Delta n$ = 2.060 and $d\Delta n$ = 0.921 calculated on the basis of Eq. (1).

In Fig. (**2**), one can see that:

- at the wavelength of λ = 1064 nm the first transmission maximum T of a passive TN effect (k = 1) occurs at $d\Delta n = \lambda \sqrt{3} / 2 = 0.921$,
- at the wavelength of λ = 1064 nm the second transmission maximum T of a TN effect (k = 2) occurs at $d\Delta n$ = 2.060.

It can be noted that the width of the second maximum of a TN effect is more narrow than the first-order one.

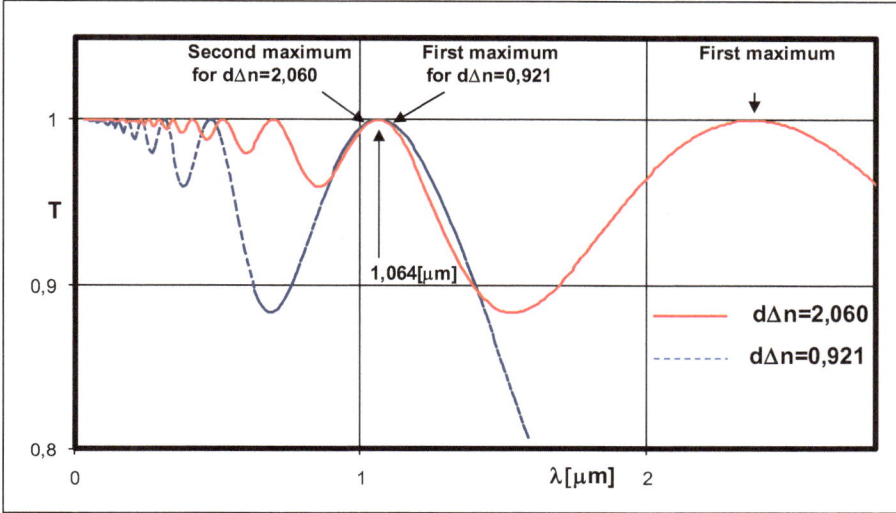

Fig. (2). The dispersion of the transmission T of a passive TN mode calculated from Eq. (1) for *dΔn* = *2.060* and *dΔn* = *0.921*.

3.2. Switching-on Time and Switching-off Time of TN Effect

Switching-on time τ_{on} and switching-off time τ_{off} of TN structure, under the driving voltage U, are defined by the following equations [5, 6]:

$$\tau_{ON} \propto \frac{\gamma d^2}{\varepsilon_o \Delta\varepsilon U^2 - \pi^2 K_{TN}}, \ \tau_{OFF} \propto \frac{\gamma d^2}{\pi^2 K_{TN}}, \quad \text{(3a, 3b)}$$

where ε_o is the permittivity of the free space and γ is the rotational viscosity of NLC. The reduced elastic constant K_{TN} (responsible for the transition from the twisted TN structure to the homeotropic HT one) and its relation with three elastic constants of pure deformations: splay (K_{11}), twist (K_{22}) and bend (K_{22}), are described in Chapter 3.

Switching-off time τ_{off} is usually much longer than switching-on time τ_{on}. Therefore, the frame time $t = \tau_{on} + \tau_{off}$ of one cycle of switching on and off of TN effect (which is one of the most critical parameters for LCC) can be estimated in the first approximation as:

$$t \approx \tau_{OFF} \propto \frac{\gamma d^2}{K_{TN}} \tag{4}$$

From Eq. (4), it results that the switching time t of LCC transducer is determined mainly by the physical properties of the NLC applied for TN effect (γ, K_{TN}) and by the cell gap d.

Since the $d\Delta n$ factor for a given NCL is strictly fixed, Eq. (4) takes the form:

$$t \propto \frac{\gamma}{(\Delta n)^2 K_{TN}} = \frac{1}{FoM}, \tag{5}$$

where the "technological factor" called *Figure of Merit* (FoM) is given by the equation:

$$FoM = \frac{(\Delta n)^2 K_{TN}}{\gamma}. \tag{6}$$

FoM describes the suitability of a given NLC for a specific application of the TN effect. From Eq. (6) it result that the faster switching dynamics is for higher FoM.

4. NEMATIC LIQUID CRYSTAL MIXTURE FOR LCC WITH TN EFFECT

According to the above considerations, an idealized LCC (neither absorbing nor reflecting the light) with transmission T higher than 95% can be developed using the TN mode working both at the first ($k = 1$) and at the second ($k = 2$) interference maximum. From the technological point of view the second maximum with $k = 2$ (at $d\Delta n = 2.060 \pm 0.200$) seems to be better than the first one $k = 1$ (at $d\Delta n = 0.921 \pm 0.100$) because the transmission higher than $T = 98\%$ is kept in the range much wider than for $k = 1$ (Fig. **3**).

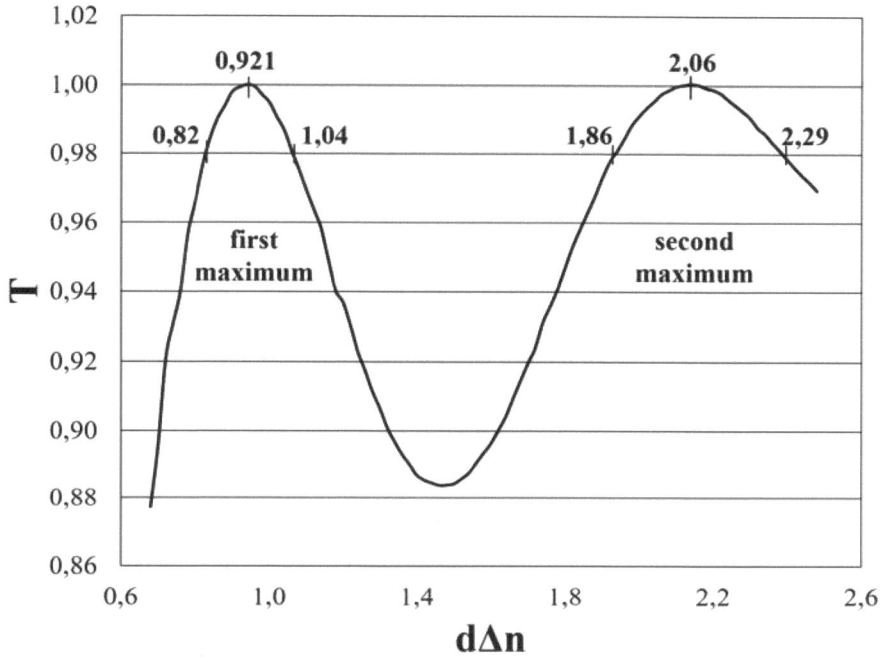

Fig. (3). The transmission *T* of the passive TN mode as a function of the factor *dΔn* calculated from Eq. (1) at $\lambda = 1064$ nm.

Switching-on time τ_{on} and switching-off time τ_{off} of a TN effect under driving with a voltage *U* can be decreased not only by decreasing both, the rotational viscosity γ and the thickness *d* of a twisted nematic layer, but also through increasing the permittivity anisotropy $\Delta\varepsilon$ and the elastic constant K_{TN}.

In order to increase the permittivity anisotropy $\Delta\varepsilon$ of a working nematic mixture, one should increase the fraction of polar components in the MNLC mixture. As a polar component of the MNLC mixture, one can regard a mesogenic or no mesogenic (but soluble in the mixture) compound with a large molecular dipole moment μ, parallel to the long molecular axis and/or with a pronounced longitudinal component of the molecular polarizability tensor. Mixtures with a high positive value of permittivity anisotropy $\Delta\varepsilon$ usually have higher rotational viscosity γ. Thus, while composing an MNLC mixture for LCC we compromised increasing of $\Delta\varepsilon$ which was inevitably accompanied by an increase of the rotational viscosity γ.

For a given wavelength of $\lambda = 1064$ nm, the *dΔn* factor has to be persistent at the value of 0.921 for the first or at 2.060 for the second interference maximum. In such a case, in order to meet the operating LCC requirements, one should develop and make an MNLC mixture with the highest possible optical anisotropy Δn. This is in order to decrease the cell gap d of an LCC transducer what leads to decrease of both; the switching-on τ_{on} and switching-off τ_{off} times.

Table 1. Material parameters of the W1825 mixture dedicated to LCC (results obtained at 25°C).

Parameter	Value, Unit
Optical anisotropy Δn (@ $\lambda = 589$ nm)	0.42
Optical anisotropy Δn (@ $\lambda = 1064$ nm)	0.37
Ordinary refractive index n_o (@ $\lambda = 1064$ nm)	1.53
Dielectric anisotropy $\Delta\varepsilon$ (@ $f = 1.5$ Hz)	17.0
Perpendicular component of permittivity tensor $\varepsilon_\perp$	4.7
Splay elastic constant K_{11}, (U_{th})	12.5 pN ($U_{th} = 0.81$ V)
Reduced elastic constant K_{TN}, (U_{TN})	16.8 pN ($U_{TN} = 1.05$ V)
Rotational viscosity γ	284 mPa·s
FoM = $K_{TN}(\Delta n)^2/\gamma$	$6.8 \cdot 10^{-12}$ m^2/s

Taking the above statements as well as the technological limitations into account, the cell gap $d = 2.5$ μm for LCC transducer was chosen. Fixing the cell gap at this particular value forced developing a dedicated MNLC mixture for TN effect, which guarantees operation at the TN mode first interference maximum at the wavelength of $\lambda = 1064$ nm. According to the Eq. (2) the optical anisotropy should be $\Delta n = 0.37$ in order to meet the condition: $d\Delta n \approx 0.92$.

According to the above assumptions, a team of chemists from the Chemistry Institute of the MUT composed the mixture W1825 with the following phase sequence: **Cr** -12.0°C **N** 136.0°C **I**. It was examined thoroughly at the Institute of Applied Physics of the MUT [7].

The detailed results of the W1825 mixture's tests were presented in Chapter 3. The chosen parameters of this mixture useful for LCC are shown in Table **1**. Since the temperature range of nematic phase of the W1825 mixture (which exists from -12.0°C to 136.0°C) is very broad in relation to operation and storage temperature ranges), the material parameters of the W1825 mixture can be regarded as near stable at LCC'S operation temperature range of -10°C to +50°C.

5. CHOSEN ELEMENTS OF THE THEORY OF LIGHT PROPAGATION IN TRANSPARENT ISOTROPIC MEDIA

5.1. Anti-reflective and Optically Matching Functional Layers

Let us consider a transparent isotropic layer TF2 of a refractive index n_2, which on one side is surrounded with a medium M1, characterized by refractive index n_1, and on the other side with medium M3, with n_3. Let us assume that TF2 is a layer of thickness d parallel to the Oxy plane of $Oxyz$ Cartesian system coordinates (see Fig. **4**).

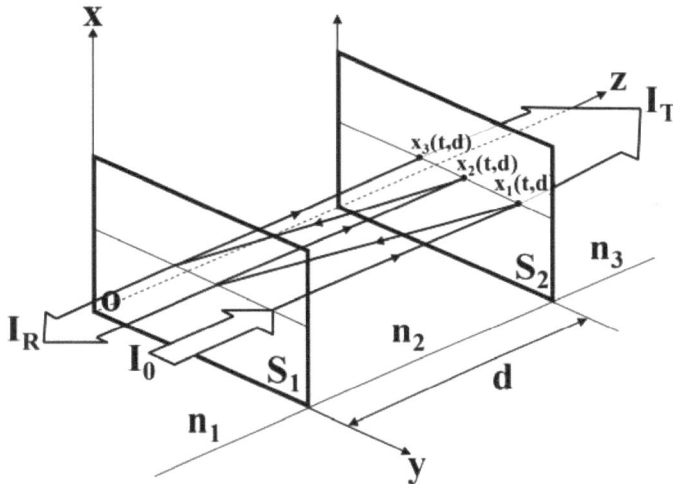

Fig. (4). The diagram of light beams' interference during light transmission through a thin plane-parallel TF2 slab of refractive index n_2 and thickness d, surrounded with isotropic media of refractive indices n_1 and n_3.

Let us assume that coherent light of intensity I_0, angular frequency ω and wavelength λ (in vacuum) incident normally on a "thin" TF2 slab from the M1 side, along the Oz axis. The E(t,z) vector of the electric field a light wave can be written in the following form:

$$E_0(t,z) = a e^{-i(\omega t - \frac{2\pi n_1 z}{\lambda})} = a e^{-i\omega t} e^{i\frac{2\pi n_1 z}{\lambda}} = a e^{-i\varphi}, \qquad (7)$$

where a is an amplitude, $\varphi = \omega t - \dfrac{2\pi n_1 z}{\lambda}$ is a wave phase of the vector $\boldsymbol{E}$. This light beam travels in the medium M1 of refractive index n_1 along the Oz axis and meets TF2 with an intensity of:

$$I_O = \frac{n_1 c}{4\pi} a^2,$$
(8)

where c is the light velocity in vacuum. The beam is partially reflected from the layer TF2 ($I_R = RI_O$), partially absorbed in the layer TF2 ($I_A = AI_O$) and the remaining part of the beam of intensity $I_T = TI_O$ goes to the medium M3 (the back reflection on S_2 is disregarded due to a low refractive index gradient). The total indices of reflection R, transmission T and absorption A describe the intensities of the waves reflected from TF2 (I_R), transmitted through TF2 to M3 (I_T) and absorbed by TF2 (I_A).

When the M1, TF2, and M3 media are dielectric ones they do not absorb the light wave, hence, ratios of the amplitudes of waves reflected on the boundaries between media with n_1 and n_2, n_2 and n_1, as well as n_2 and n_3, to the amplitude of falling wave, r_{12}, r_{21} and r_{23}, respectively can be written as follows [4, 8 - 10]:

$$r_{12} = \frac{n_1 - n_2}{n_1 + n_2}, \ r_{21} = \frac{n_2 - n_1}{n_1 + n_2} \text{ and } r_{23} = \frac{n_2 - n_3}{n_2 + n_3}.$$
(9)

The amplitude ratios t_{12}, t_{21} and t_{23} of passing waves to falling ones on the same boundaries between media with n_1 and n_2, n_2 and n_1, as well as n_2 and n_3, can be expressed as [4, 8 - 10]:

$$t_{12} = \frac{2n_1}{n_1 + n_2}, \ t_{21} = \frac{2n_2}{n_1 + n_2} \text{ and } t_{23} = \frac{2n_3}{n_2 + n_3}$$
(10)

Therefore, vectors E_R of the reflected wave (from TF2 layer) and E_T, of the passing one, can be considered as the sum of many component waves with decreasing amplitudes caused by multiple reflections on two TF2 layer boundaries. For the given moment t and at the position $z = 0$, the total light wave $E_R(t, 0)$ reflected from the TF2 layer is:

$$E_R(t,0) = r_{12} a e^{-i\omega t} + t_{12} r_{23} t_{21} a e^{-i\omega t} e^{i\Delta} + t_{12} r_{23} r_{21} r_{23} t_{21} a e^{-i\omega t} e^{i2\Delta} + t_{12} r_{23} (r_{21} r_{23})^2 t_{21} a e^{-i\omega t} e^{i2\Delta} + ...$$

Knowing that each of components of the light wave written above differs from its antecedent, due to double passage across the TF2 layer and two reflections on boundaries, the phase difference Δ between these component waves can be written as follows:

$$\Delta = \frac{4\pi n_2 d}{\lambda} + \varepsilon_{21} + \varepsilon_{23},$$

where ε_{21} and ε_{23} are the phase shifts (increments) at the reflection from the boundary between M1 and TF2 as well as between TF2 and M3, respectively.

When the media M1, TF2 and M3 are dielectric ones then $\varepsilon_{21} = \varepsilon_{23} = \pi$ [11] and:

$$\Delta = \frac{4\pi n_2 d}{\lambda} + 2\pi \qquad (11)$$

In this case, the complex amplitude A_R of the wave $E_R(t,0) = A_R e^{-i\omega t}$ reflected from TF2 has a form of:

$$A_R = \frac{r_{12} - (r_{12}r_{21}r_{23} - t_{12}r_{23}t_{21})e^{i\Delta}}{1 - r_{21}r_{23}e^{i\Delta}} \qquad (12)$$

Knowing the complex amplitude A_R and its conjugate amplitude $A_R{}^*$ of the light wave reflected from TF2 the intensity I_R of such wave can be calculated as:

$$I_R = \frac{n_1 c}{4\pi} A_R A_R^* = \frac{n_1 c}{4\pi} a^2 \frac{r_{12}^2 - 2(r_{12}r_{21}r_{23} - t_{12}r_{23}t_{21})r_{12}\cos\Delta + (r_{12}r_{21}r_{23} - t_{12}r_{23}t_{21})^2}{(1 - r_{21}r_{23}e^{i\Delta})(1 - r_{21}r_{23}e^{-i\Delta})} = RI_O \quad (13)$$

For the given moment t and the position $z = d$ the total light wave $E_T(t, d)$ passing through TF2 layer can be written as:

$$E_T(t,d) = t_{12}t_{23}ae^{-i\omega t} + t_{12}r_{23}r_{21}t_{23}ae^{-i\omega t}e^{i\Delta} + t_{12}(r_{23}r_{21})^2 t_{23}ae^{-i\omega t}e^{i2\Delta} + t_{12}(r_{23}r_{21})^3 t_{21}ae^{-i\omega t}e^{i3\Delta} + \dots$$

The right-hand side of the above formula is the geometric series with the quotient:

$r_{23}r_{21}e^{i\Delta}$ so the complex resultant amplitude A_T of the light wave

$E_T(t,d) = A_T\, e^{-i\omega t}$ passing through a TF2 layer has a form of:

$$A_T = t_{12}t_{23}a(1 + r_{23}r_{21}e^{i\Delta} + (r_{23}r_{21})^2\, e^{i2\Delta} + (r_{23}r_{21})^3\, e^{i3\Delta} + \ldots \quad) = \frac{t_{12}t_{23}a}{1 - r_{23}r_{21}e^{i\Delta}}.$$

Knowing A_T, and $A_T{}^*$ one can determine the intensity I_T of a light wave passing through a TF2 layer as:

$$I_T = \frac{n_3 c}{4\pi} A_T A_T^* = \frac{n_3 c}{4\pi} a^2 \frac{t_{12}^2 t_{23}^2}{1 + r_{23}^2 r_{21}^2 - 2r_{23}r_{21}\cos\Delta} = TI_O. \tag{14}$$

While analyzing Eqs. (13) and (14) one can note several observations.

a) When the relation:

$$n_2 d = \lambda/4 \text{ and } n_2 = \sqrt{n_1 n_3}, \tag{15}$$

are satisfied with the Eq. (13) achieves its minimum ($I_R = 0$, $R = 0$) while Eq. (14) achieves the maximum value ($I_T = I_O$, $T = 1$). It means that, a thin, transparent TF layer surrounded by the media M1 and M2 for which d and n meet the condition from Eq. (15) can;

- decrease (interferential) light reflection (R) coefficient on the boundary between medium M1 and TF2 deposited on M3; in this case, such a fine TF2 layer acts as an antireflective layer;
- increase (interferential) light transmission (T) coefficient from the medium M1 to M3 through TF2; in this case, thus TF2 selected layer acts as a matching layer.

b) Let us consider a dielectric, thin TF layer of the refractive index n_2 surrounded by media with a high refractive indices $n_1 = n_3 = n$, or by partially transparent conductive layers (ITO). At the conductive surface, the relation:

$$\tau + \rho + \alpha = 1 \tag{16}$$

is satisfied for each light beam falling on this surface from the side of M1 or M3, where α is the absorption coefficient, τ is the transmission coefficient of the form:

$$\tau = (t_{12})^2 = (t_{23})^2 = (\frac{2n}{n+n_2})^2, \tag{17}$$

and ρ is the reflection coefficient

$$\rho = r_{21}^2 = r_{23}^2 = (\frac{n_2-n}{n+n_2})^2. \tag{18}$$

After substitution of (17) and (18) to (14) one gets:

$$I_T = \frac{nc}{4\pi}a^2\frac{\tau^2}{1+\rho^2-2\rho\cos\Delta} = I_0\frac{\tau^2}{(1-\rho)^2}\frac{1}{1+\frac{4\rho}{(1-\rho)^2}sin^2(\frac{1}{2}\Delta)}. \tag{19}$$

According to Eq. (16), based on the law of energy conservation and applying an F coefficient of "finesse" defined as [12]:

$$F = \frac{4\rho}{(1-\rho)^2}, \tag{20}$$

the Eq. (19) can be written:

$$I_T = (1-\frac{\alpha}{(1-\rho)})^2\frac{1}{1+F sin^2(\frac{1}{2}\Delta)}I_0. \tag{21}$$

Eq. (21) describes the total intensity I_T of light passing through the absorbing ($\alpha \neq 0$) and reflecting ($\rho \neq 0$) TF layer [4, 8 - 10].

5.2. Total Ratio of Transmission (T) and of Reflections (R) of the Plane Parallel Dielectric Plate

Let us consider a transparent, isotropic layer (again see Fig. **4**) of refractive index n_2 surrounded by air ($n = n_1 = n_3 = 1$). Let the thickness of this plate be high enough, so no interference effects of the TF layer will show up. When the plate is the dielectric one it does not absorb the light ($\alpha = 0$) and then coefficients of reflection ρ and transmission τ can be written as:

$$\rho = (\frac{n_2 - 1}{n_2 + 1})^2 , \ \tau = 1 - \rho = (\frac{2n_2}{n_2 + 1})^2 . \tag{22}$$

When monochromatic light of intensity I_0 enters this plate normally the total coefficients of transmission T and reflection R are appropriate amounts of all reflection and transmission effects occurring on two-side surfaces of a dielectric plate:

$$T = (1-\rho)^2 + \rho^2(1-\rho)^2 + \rho^4(1-\rho)^2 + ... \quad = \frac{1-\rho}{1+\rho}, \ R = 1 - T = \frac{2\rho}{1+\rho} . \tag{23}$$

6. LCC FOR A LASER RANGEFINDER

6.1. TN Effect with MNLC Mixture Selected for LCC

In order to determine static, dynamic and spectral characteristics of TN effect with the W1825 mixture (see Chapter 3) four types of special measuring cells (TN10, TN70, TN100, and TN500) of cell gap $d \in$ (1.5 µm, 3.5 µm) were manufactured at the MUT (see Chapter 3).

All empty cells selected for this purpose were flat (no interference fringes of equal inclination observed) and the whole observed surface was of the same color in a fluorescent bulb light.

Figs. (**5** and **6**) present the results of measurements of the dispersion of the transmission $T(\lambda)$ of a two (2.5TN10 and 2.6TN500) cells filled with W1825 mixture ($\Delta n = 0.37$ at $\lambda = 1064$ nm, at 25°C) placed between crossed polarizers. The dispersion of the transmission $T(\lambda)$ (see Fig. **5**) was measured using JASCO V670 spectrometer at the polarization plane of incoming light parallel to the director **n** at the incident site of the TN slab. The dispersion of the transmission $T(\lambda)$ presented in Fig. (**6**) was obtained at the polarization plane perpendicular to the director **n** at the incident site of the TN slab.

2.5 TN 10 cell with LCM

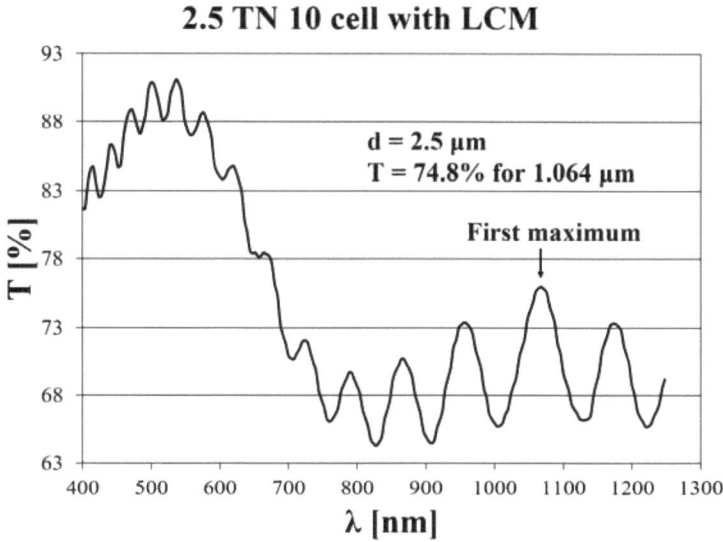

Fig. (5). The dispersion of the transmission T(λ) of the 2.5TN10 cell filled with W1825 mixture (of $\Delta n = 0.37$ at $\lambda = 1064$ nm, at 25°C) placed between crossed polarizers.

2.5 TN 500 cell with LCM

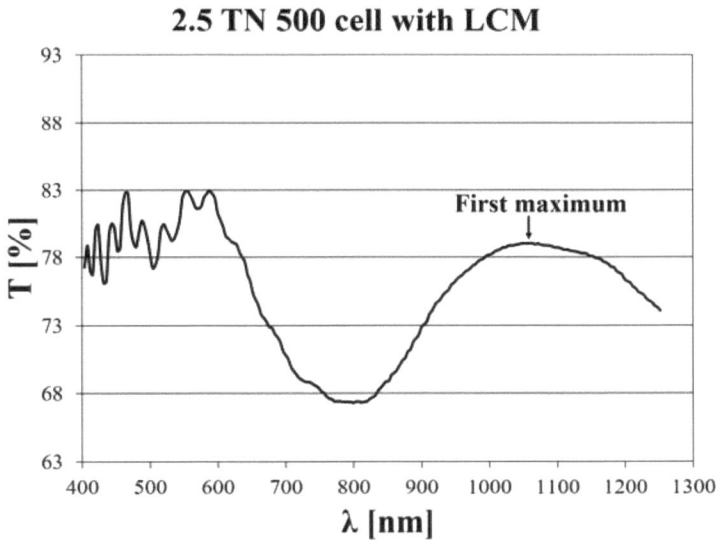

Fig. (6). The dispersion of the transmission T(λ) of a 2.6TN500 cell filled with W1825 mixture (of $\Delta n = 0.37$ at $\lambda = 1064$ nm, at 25°C) placed between crossed polarizers.

Switching-on τ_{on} and switching-off τ_{off} times observed in the 2.6TN500 cell filled with W1825 at different driving voltages U are gathered in Table **2**.

Table 2. Switching-on τ_{on} and switching-off τ_{off} times observed in a 2.6TN500 cell filled with W1825 at different driving voltages.

Voltage U	Switching-on times τ_{on} [ms]	Switching-off times τ_{off} [ms]
5 V	2.17	5.85
10 V	0.67	6.03
15 V	0.45	6.33
20 V	0.33	6.40

As one can see in Figs. (**5**) and (**6**) spectral positions of the maximum transmissions T at the first maxima of the positive TN effect observed for both cells 2.5TN10 and 2.6TN500 of almost the same cell gap filled with W1825, occur for almost the same wavelength of $\lambda = 1064$ nm. Moreover, switching-on τ_{on} and switching-off τ_{off} times in the 2.6TN500 cell meet the requirements for LCC (see Table **1**). The above indicates that the W1825 mixture promise to build LCC for a space laser rangefinder.

Despite the fact that the factor $d\Delta n = 0.37 \times 2.5$ (or 2.6) ≈ 0.92 for the W1825 mixture matches the first interference maxima of passive TN effects in those cells very well, the overall transmissions T at wavelength $\lambda = 1064$ nm in the 2.5TN10 and 2.6TN500 cells filled with W1825 achieve barely the values of 74.8% and 79.1% (see Figs. **5** and **6**). Such values are far from meeting technical requirements for the transmission T ($T > 95\%$) in LCC. Moreover, in Fig. (**5**) one can see distinct "thickness interference fringes" at the main TN interference curve of the transmission T registered in the 2.5TN10 cell with the W1825 mixture. These sharply outlined "thickness" fringes come from multiple reflections of a light beam in a plane-parallel slab of MNLC mixture bound by conductive ITO electrodes. The undesirable phenomenon of "thickness fringes" creation can be described, in the first approximation, by Eq. (21).

In the case of the transmission T shown in Fig. (**5**) a mismatch of refractive indices $n \sim 1.95$ [13] of an ITO layer (with a typical density, so-called "dense ITO", the thickness of ~ 25 nm and resistance of ~ 10 $\Omega/\square$) with the refractive index n of the W1825 mixture ($n_o \sim 1.54$) according to Eqs. (18) and (20), causes the increase of $\rho \sim 0.014$ and $F \sim 0.06$. As a result, modulation of the transmission T the wavelength λ domain of is clearly visible.

Taking the above into account as well as knowing that losses are caused by:

a. absorption in a layer of LC mixture W1825,
b. absorption in polymer alignment layers of a TN cell,
c. light diffusion in all TN cell's layers and on the boundaries between those

layers,

d. light absorption in conductive ITO electrodes (see Figs **7** and **8**),
e. light absorption in glass plates (see Fig. **9**),
f. high reflections on all interfaces (air-glass-ITO-MNLC-ITO-glass-air) of a filled TN cell (see Figs **7-9**),

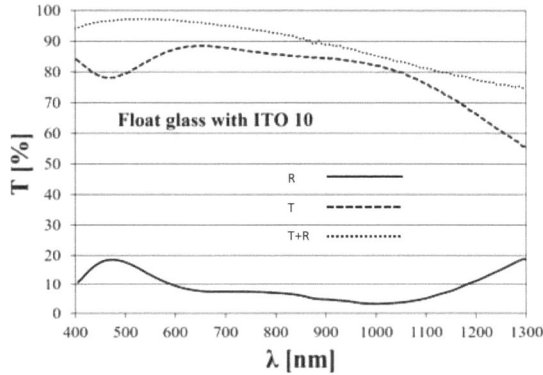

Fig. (7). The dispersion of the transmission T(λ) and reflection $R(\lambda)$ at plane parallel float glass plate of a thickness of 1.1 mm coated by an ITO layer of a specific resistivity of ρ=10 Ω/$\square$.

it should be noted that it is impossible to build an LCC of the transmission $T > 95\%$ based on a simple glass 2.5TN500 cell made of commercial "float" glass with a commercial ITO layer with the highest, commercially available, the resistance of 500 Ω/$\square$, filled with the W1825 mixture.

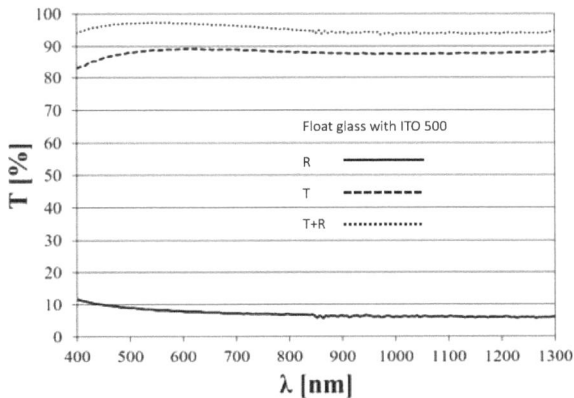

Fig. (8). The dispersion of the transmission T(λ) and reflection $R(\lambda)$ at plane parallel float glass plate of a thickness of 1.1 mm coated by an ITO layer of a specific resistivity of ρ=500 Ω/$\square$.

6.2. The Optically Matched LCC Transducer with W1825 Mixture

When one wants to increase the transmission T of LCC above 95%, one has to minimize, and sometimes to eliminate, mentioned above, causes of "light losses" at the transition of a laser beam.

Our earlier research on working NLCs and LCDs prove that losses resulting from light absorbing by few micrometer LC layers or tens of nanometer polymer alignment layers are minor. NLC slab of thickness $d = 45$ µm usually exhibits absorption $A < 5\%$, while the absorption of a 3 µm slab of the W1825 mixture was unmeasurable. A polymer alignment layer (PL) of thickness $d = 15$ nm exhibits an absorption $A < 1\%$, while the absorption of a 30 nm layer of a NISSAN SE-130 polymer applied for LCC was unmeasurable.

In order to reduce the absorption A of LCC, Quartz Plates (QP) of thickness $d = 1.5$ mm and $n = 1.54$ (at $\lambda = 1064$ nm [14]) was used instead of a common glass plate. Accordingly to Eqs. (22) and (23) reflection for these silica windows is $\rho = 0.034$ and theoretical values of the transmittance and the reflectance are $T = 93.5\%$ and $R = 6.5\%$, respectively. Since $T+R=1$, practically, these silica windows do not absorb light within the VIS and NIR ranges (see Fig. **10**).

In order to minimize the light scattering on the boundaries of QPs, both sides of QPs were mechanically polished as to achieve the best possible optical quality. After that, the flatness of QP was better than $\lambda/4$ (at $\lambda=633$ nm), and the wedge-shaped character of QPs was smaller than 8.

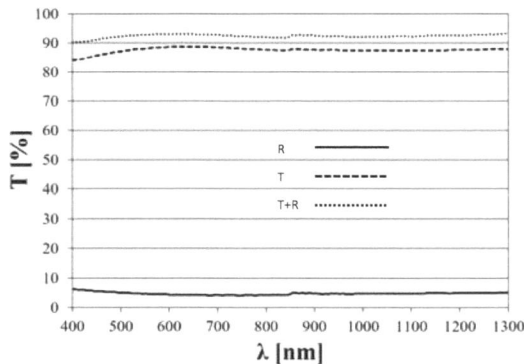

Fig. (9). The dispersion of the transmission T(λ) and reflection R(λ) of a plane parallel mechanically polished float glass plate of thickness 1.1 mm and $n = 1.52$ applied as a substrate for 2.5TN10 and 2.5TN500 cells.

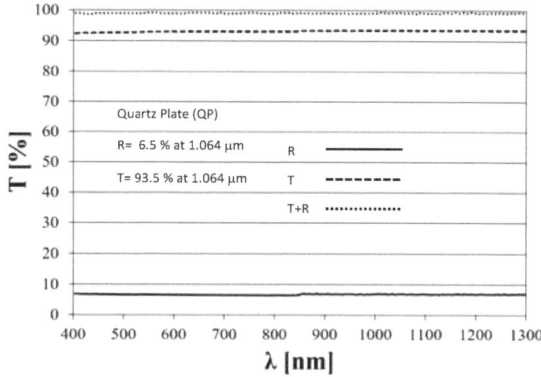

Fig. (10). The dispersion of the transmission T(λ) and reflection $R(\lambda)$ of a plane parallel mechanically polished "Quartz Plate" (QP) of thickness 1.5 mm and $n = 1.54$ applied as a substrate for LCC.

In order to lower optical losses caused by light absorption by conductive ITO layers as well as the Fresnel losses caused by interfaces between ITO layers and QP substrates, it was decided to deposit thinner Porous ITO (PITO) layers on QP substrates by using vacuum evaporation technique [13]. From our earlier investigations, we know that the ITO layer of thickness from 15 nm (for 500 $\Omega/\square$) to 180 nm (for 10 $\Omega/\square$) is characterized by a refractive index n value close to the extraordinary refractive index n_e of the W1825 mixture at $\lambda = 1064$ nm. In Fig. (**6**) we cannot see "thickness" fringes in the area of $\lambda = 1064$ nm, which indicates that n_e of the W1825 mixture at $\lambda = 1064$ nm is (approximately) the same as the coefficient of ITO layers covered with SE-130. In order to determine the optimal conditions for the vacuum deposition of the PITO layers needed for LCC, three PITO layers were deposited on the identical three substrates (QP). As a result of variation of the thicknesses of deposited layers, as well as of other deposition parameters, three samples of PITO layers with sheet resistances 400 $\Omega/\square$, 1000 $\Omega/\square$ and 1200 $\Omega/\square$ were obtained. Fig. (**11**) presents the results of the dispersion of the transmission T of PITO layers of different specific resistance. One can see clearly in Fig. (**11**) that transmittance ($T_{1200} = 93.1\%$ at $\lambda = 1064$ nm) of the quartz substrate QP with the deposited layer PITO1200 (with a sheet resistance of 1200 $\Omega/\square$) is almost the same as transmission ($T = 93.5\%$ for $\lambda = 1064$ nm) for the bare QP substrate. It means that by applying evaporated PITO1200 layers, one can significantly lower the absorption, as well as a refractive index of the conducting layer.

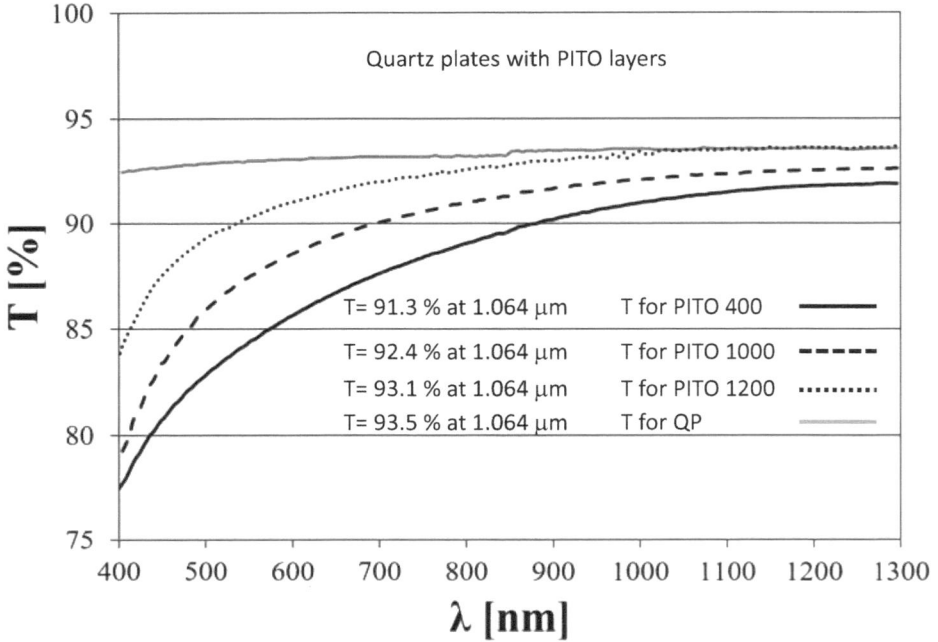

Fig. (11). The dispersion of the transmittance $T(\lambda)$ for: bare QP, QP with PITO1200 layer (1200 $\Omega/\square$), QP with PITO1000 layer (1000 $\Omega/\square$), QP with PITO400 layer (400 $\Omega/\square$).

In order to avoid direct contact between both transparent, conductive PITO electrodes of relatively large surface (LCC aperture cannot be lower than 15 mm) with a working mixture (W1825 with $n_o = 1.53$ at $\lambda = 1064$ nm), a thin SiO_2 blocking film (BF) was evaporated on PITO electrode. Since the refractive indices of PITO electrode ($n = 1.54$) and that of BF ($n = 1.54$) is almost the same as the ordinary refractive index ($n_o = 1.53$) of the W1825 mixture at $\lambda = 1064$ nm the BF thickness d should match the one of those given by Eq. (15). If the thickness of the blocking layer BF is $d = \lambda/(4n)$, the Fresnel reflections cannot be seen on the boundary between PITO ($n = 1.54$) and BF ($n \sim 1.54$), as well as on the next one; between BF ($n = 1.54$) and MNLC mixture ($n_o \sim 1.53$).

In order to remove (minimize) the relatively high reflections R on LCC external surface or on the boundaries of QP with air, these surfaces were deposited with anti-reflective (AR) coatings. AR coatings were double, standard Al_2O_3-MgF_3 layers of thicknesses matching with $\lambda = 1064$ nm.

The transmittance T (measured with a JASCO V670 spectrometer) of a QP coated with an AR layer on the one part and covered with layers of PITO, BF and PL on the other part, is presented in Fig. (**12**). The obtained transmission $T > 96\%$ at $\lambda = 1064$ nm was confirmed by authors by means of direct measurements of energy losses of the laser beam of $\lambda = 1064$ nm passing through QP with AR, ITO, BF, and PL layers.

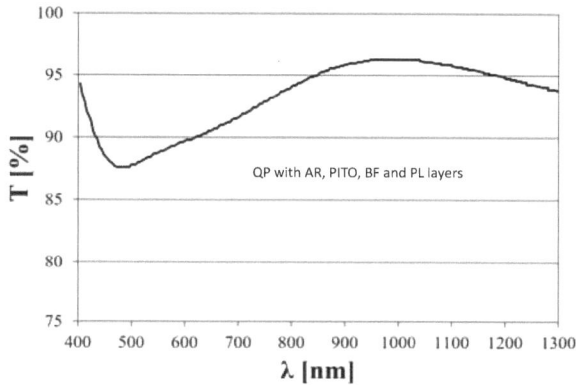

Fig. (12). Transmissions $T(\lambda)$ for QP plates with: AR (anti-reflective layer of Al_2O_3-MgF_3), PITO, BF (SiO_2 layer) and PL (SE-130 polyimide).

Taking the above into consideration, an optically matched LCC of $T > 95\%$ should be constructed as a multilayer structure presented in Fig. (**13**).

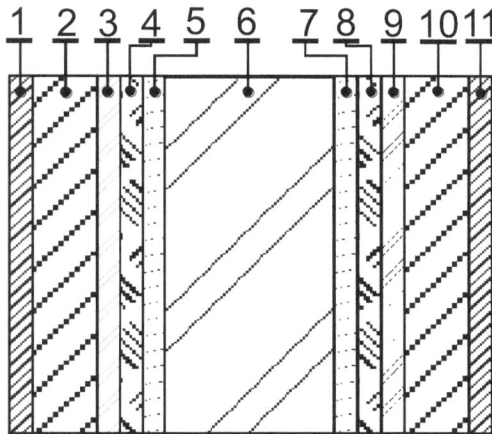

Fig. (13). Cross-section of an optically matched LCC: 1, 11 - dielectric, anti-reflective (AR) layer; 2 and 10 - quartz plates (QP); 3 and 9 - PITO transparent electrodes; 4 and 8 - SiO_2 (BF) blocking layers; 5 and 7 - SE-130 alignment layers (PL); 6 - working liquid crystal mixture W1825.

In order to fabricate developed LCC:

- the QP polished on both sides at the precision of $\lambda/4$ @ 633 nm,
- with dielectric anti-reflective AR layers deposited on one side and
- with transparent PITO electrodes on the other side,
- protected by dielectric blocking layers BF,
- with rubbed alignment PL layers,
- as well as with contact areas for soldering of control cables made of evaporated gold-on-chrome layers,

were sealed using thermo-cured adhesive.

The layout of the LCC cell is shown in Fig. (**14**). In order to determine the desired thickness $d = 2.5$ μm the glass spacers of a diameter of 2.5 μm were added to the sealing glue. The example of dispersion of the transmission $T(\lambda)$, measured using a JASCO V670 spectrometer of an empty LCC cell with all functional layers (AR, PITO, BF, PL), is presented in Fig. (**15**).

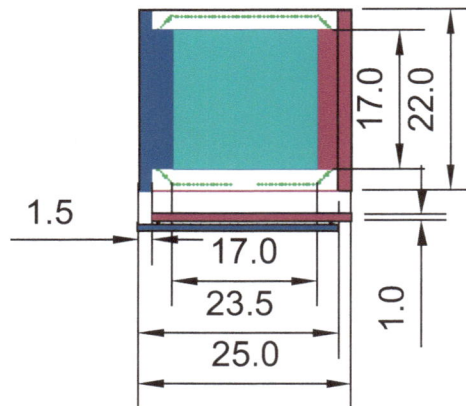

Fig. (14). The layout of an optically matched LCC.

Empty LCC was filled with a working W1825 mixture described above. The filling process took place in a vacuum chamber and then the LCC cell was encapsulated at 30°C using an optical adhesive OK-15 dedicated for use in space.

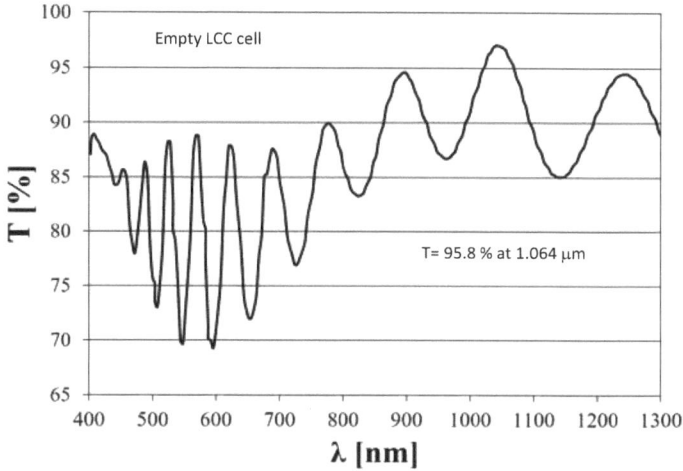

Fig. (15). The dispersion of the transmission T(λ) of an empty LCC with all (AR, PITO, BF, PL) functional layers.

6.3. Tests of LCC with the Liquid Crystal Mixture W1825

All LCCs were developed and manufactured at MUT laboratories (in 4 series of 12 items), LCCs were thoroughly investigated and tested at MUT laboratories in Warsaw, Poland. The following characteristics of LCC were investigated at MUT:

- Transmittance $T(\lambda)$,
- Reflectance $R(\lambda)$,
- Absorption $A(\lambda)$,
- Switching-on time τ_{on},
- Switching-off time τ_{off},
- Contrast ratio CR at $\lambda = 1064$ nm,
- Systematic aging research was also done,
- Limited studies concerned the work of LCC in vacuum conditions, as well as a test while rapid temperature changes.

In order to illustrate the process of LCC tailoring and testing of the high transmittance regime ($T > 95\%$) let us present two charts illustrating the dispersion of the transmittance T of the same LCC with cell gap of $d = 2.3$ μm, filled with W1825 mixture, placed between a two crossed polarizers and measured by means of JASCO V670 spectrometer. When the polarization plane of light incident on LCC is parallel to the director $\boldsymbol{n}$ in an NCL layer at the incident side (see Fig. **16**), then the light wave experiences the extraordinary refractive index n_e of the W1825 mixture which makes LCC optically mismatched what results in $T < 95\%$. When the polarization plane of light incident LCC is perpendicular to the director $\boldsymbol{n}$ in an NLC layer at the entrance to this layer (see Fig. **17**), then light leaving PITO and PL with $n \sim 1.54$ experiences $n_o \sim 1.53$ of the W1825 mixture. The difference of refractive indices is smaller, which makes the LCC optically matched and yields $T > 95\%$.

The results of LCC studies from series 1 performed at the MUT (based on commercial flat glass with ITO 500 $\Omega/\square$ with AR and PL layers) were promising. LCC with W1825 mixture obtained at MUT, and tested at the MUT and in Russian Federation indicated $T \sim 89\%$ and were definitely faster ($\tau_{on} < 1$ ms, $\tau_{off} < 8$ ms at 10 V) than those previously considered for the use in a space mission. Russian LCC (developed and manufactured in the LCD Plant in Saratov, Russian Federation, based on silica with deposited AR, ITO and PL layers and containing a Chinese commercial liquid crystal mixture) hardly achieved $T \sim 84\%$ and acceptable switching-on and off times, but their driving voltages U were twice as high as Polish ones ($\tau_{on} < 1.5$ ms, $\tau_{off} < 10$ ms at 20 V_{RMS}). Along with acquiring knowledge and experience in the area of LCC for space applications technology procedures of "anti-reflective", "blocking", "matching" and alignment (AR, PITO, BF and PL) layers deposition, the obtained LCC were improving with each series.

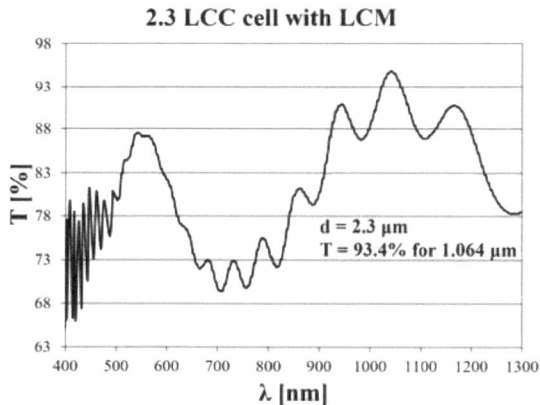

Fig. (16). The dispersion of the transmission T(λ) of 2.3LCC ($d = 2.3$ µm) filled with W1825 ($\Delta n = 0.37$ at $\lambda = 1064$ nm). $T(\lambda)$ was obtained using a JASCO V670 spectrometer in which 2.3LCC was placed between two crossed polarizers and polarization plane of light falling on LCC was parallel to the director $\boldsymbol{n}$ in NCL at the entrance to TN layer.

2.3 LCC cell with LCM

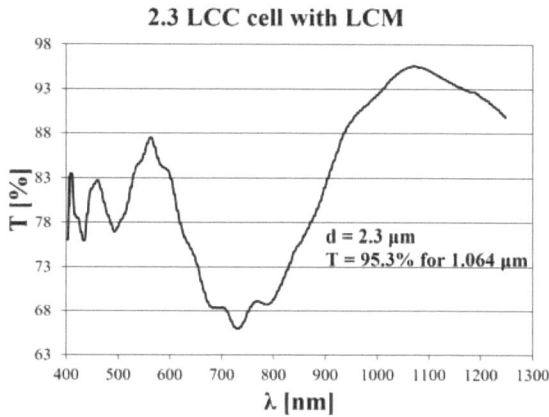

Fig. (17). The dispersion of the transmission T(λ) of 2.3LCC ($d = 2.3$ µm) filled W1825 ($\Delta n = 0.37$ at $\lambda = 1064$ nm). $T(\lambda)$ was obtained using a JASCO V670 spectrometer in which 2.3LCC was placed between two crossed polarizers and polarization plane of light falling on LCC was perpendicular to the director n in NCL at the entrance to TN layer.

In series 2 (based on QP with AR, PITO and PL) some laboratory LCC models with $T \sim 95\%$ were happily obtained, but objections to optical purity of some LCC still raised.

LCC from series 3 (in which all optical layers developed at MUT: AR, PITO, BF and PL as well as fully professional materials for LCD sticking and caulking were applied) met all requirements. After testing of optical quality and operating parameters, each LCC manufactured at the MUT (meeting technical requirements) was given a number, *i.e.*, LCC47 (7 from series 4) and a metric (similar to the one shown below for LCC47), and, next 6 items from series 3 (with the best technical parameters) were sent for further testing to VSOI in St. Petersburg, Russia.

7. LIQUID CRYSTAL CELL FOR SPACE-BORNE LASER RANGEFINDER FOR INTERNATIONAL SPACE MISSION "PHOBOS-GROUND"

The Liquid Crystal Cell (LCC) sent to Vavilov State Optical Institute (VSOI), St. Petersburg, Russian Federation, was thoroughly studied and tested in VSOI in the presence and with the active participation of authors of this monograph. It is necessary to mention that all the results obtained at VSOI were not worse than those obtained at MUT. At VSOI, LCCs with parameters confirmed by a common, official protocol was installed in the laser rangefinders and were tested (without Polish party) at VSOI and in space fields in Russia. While tested in one of the Russian space fields some (3 out of 6) LCCs from series 3 depressurized during several days of testing under high (space) vacuum. After a thorough analysis, it turned out that what failed was the "professional" adhesive used by the Polish party (widely used in LCD technology) for the sealing of LCC. After applying OK-15 adhesive, so-called "space adhesive" developed in Russia in 2007 within a separate three-year grant, 10 pieces of LCC from series 4 met all the tests provided at MUT and, next, in Russian Federation. Due to significant delays of the "Phobos-Ground" project the Russian party guided by former space programs experience canceled the tests concerning the resistance of LCC' to long-term cosmic radiation to the total dose designed for the program.

Table 3. Collected results of LCC research (averaged across of 10 pieces of LCC of series 4) performed at MUT, and then confirmed in Russian Federation.

Test Type	Value, Unit
Transmittance (T) @ $\lambda = 1064$ nm	Not lower than 95.3%
Reflectance (R) @ $\lambda = 1064$ nm	Not higher than 2%
Absorbance (A) @ $\lambda = 1064$ nm, ($A = 100-(T+R)$)	Not higher than ~ 2%
Switching time (τ_{on}), ms (for $U = 10$ V)	Lower than 0.7 ms
Switching time (τ_{off}), ms	Lower than 7 ms
Contrast ratio (CR) for $\lambda = 1.064$ μm	Not lower than 300:1

Table 4. Technical parameters of LCC47.

Parameter	Value, Unit
External dimensions	25.0 mm×22.0 mm×3.1 mm
Working aperture	17 mm
Operating temperature	from 20° to 40°C
Storage and transportation temperature	from -10°C to 50°C
Work wavelength λ	1.064 μm
Electro-optical Effect applied in LCC	TN (Twisted Nematic)
Liquid Crystal Mixture (LCM) applied in LCC	W1825K
Optical anisotropy Δn of LCM for $\lambda = 1.064$ μm at 25°C	0.37
Ordinary refractive index n_o of LCM for $\lambda = 1.064$ μm at 25°C	1.53
Dielectric anisotropy $\Delta \varepsilon$ of LCM for $f = 1.5$ Hz at 25°C	17.0
Thickness d of LCM layer in LCC	2.5 μm
Transmission T at the wavelength 1.064 μm,	97.8%
Reflection R at the wavelength 1.064 μm,	< 1%
Absorption A at the wavelength 1.064 μm, ($A = 100-(T+R)$)	~ 2%
Switch-on time τ_{on} for Driving Voltage $U_d = 10$ V	0.65 ms

(Table 4) cont.....

Parameter	Value, Unit
Switch-off time τ_{off} for Driving Voltage $U_d = 10$ V	6.04 ms
Contrast Ratio CR in switch-on regime	Not less than 300:1

It was recognized that positive preliminary studies on resistance to cosmic radiation to a lower dose are sufficient because LCCs are not placed in the open space but in a radiation-free inside of the lander's capsule.

Collected results of LCC studies (averaged across 10 pieces of LCC of series 4) performed at MUT, and then confirmed in the Russian Federation, are gathered in Table **3**.

After meeting by LCC all the tests confirmed by special protocols, the Russian party chose two LCCs (LCC47 and LCC43) with the best technical parameters (see data for LCC47 presented in Table **4**, below) and installed them in the laser rangefinder of the space lander "Phobos-Ground" which was next launched into space November 8th, 2011, from Kazakhstan.

CONSENT FOR PUBLICATION

Not applicable.

CONFLICT OF INTEREST

The author(s) confirms that there is no conflict of interest.

ACKNOWLEDGEMENTS

Declared none.

REFERENCES

[1]   E. Nowinowski-Kruszelnicki, L. Jaroszewicz, Z. Raszewski, L. Soms, W. Piecek, P. Perkowski, J. Kędzierski, R. Dąbrowski, M. Olifierczuk, and E. Miszczyk, "Liquid crystal cell for space-borne laser rangefinder to space mission applications", *Opto-Electron. Rev.,* vol. 20, no. 4, pp. 315-322, 2012. [http://dx.doi.org/10.2478/s11772-012-0045-7]

[2]   E. Nowinowski-Kruszelnicki, and C. Rymarz, "Electrodynamic instabilities in liquid crystals according to the principle of conservation of moment of momentum", *J. Tech. Phys.,* vol. 19, no. 2, p. 21, 1978.

[3]   C.H. Gooch, and H.A. Tarry, "The Optical Properties of Twisted Nematic Liquid Crystal Structures with Twist Angle < 90 degrees", *J. Phys. D Appl. Phys.,* vol. 8, no. 13, pp. 1575-1584, 1975. [http://dx.doi.org/10.1088/0022-3727/8/13/020]

[4]   R.W. Ditchburn, *"Light"*, 22nd ed. (The Student's Physics.), London and Glasgow: Blackie and Son, Ltd., 1963.

[5]   R. Dąbrowski, J. Dziaduszek, A. Ziółek, Ł. Szczuciński, Z. Stolarz, G. Sasnouski, V. Bezborodov, W. Lapanik, S. Gauza, and S-T. Wu, "Low viscosity, high birefringence liquid crystalline compounds and mixtures", *Opto-Electron. Rev.,* vol. 15, pp. 47-51, 2007.
[http://dx.doi.org/10.2478/s11772-006-0055-4]

[6]   K. Tarumi, U. Frinkenzeller, and B. Schuler, "Dynamic behavior of twisted nematic liquid crystals", *Jpn. J. Appl. Phys.,* vol. 31, pp. 2829-2836, 1992.
[http://dx.doi.org/10.1143/JJAP.31.2829]

[7]   E. Nowinowski-Kruszelnicki, J. Kędzierski, Z. Raszewski, L. Jaroszewicz, R. Dąbrowski, W. Piecek, P. Perkowski, M. Olifierczuk, K. Garbat, M. Sutkowski, E. Miszczyk, K. Ogrodnik, P. Morawiak, M. Laska, and R. Mazur, "High birefringence liquid crystal mixtures for lc electro-optical devices", *Opt. Appl.,* vol. 42, no. 1, pp. 167-180, 2012.

[8]   M. Born, and E. Wolf, *Principles of Optics, Seventh Edition - Electromagnetic Theory of Propagation, Interference and Diffraction of Light,* Cambridge University Press: London, 2003.

[9]   *Measuring Techniques.* vol. 3. Elsevier Science: Amsterdam, 1993.M. Pluta, Advanced light microscopy, Volume 3 - Measuring Techniques

[10]   M. Pluta, "Simplified polanret system for microscopy", *Appl. Opt.,* vol. 28, no. 8, pp. 1453-1466, 1989.
[http://dx.doi.org/10.1364/AO.28.001453] [PMID: 20548681]

[11]   T. Cesarz, S. Kłosowicz, E. Nowinowski-Kruszelnicki, and J. Żmija, "Liquid crystal elements of laser optics. The optical isolator", *Mol. Cryst. Liq. Cryst. (Phila. Pa.),* vol. 193, pp. 19-23, 1990.

[12]   A. Kieżun, L.R. Jaroszewicz, A. Walczak, and E. Nowinowski-Kruszelnicki, "Direct measurement of refractive index profile in liquid crystal planar waveguides", *Proc. SPIE,* vol. 3745, pp. 92-98, 1999.
[http://dx.doi.org/10.1117/12.357766]

[13]   F.W. Xing Yon, D.J. Mont, and M.F. Poxson, "Reflactive-index-matched indium-thin–oxide for liquid crystal display", *Jpn. J. Appl. Phys.,* vol. 48, pp. 120203.1-3, 2009.

[14]   Heraeus, "Daten Und Eigenschaften, Quarzglas fur die Optic", *datasheet,* 2009.

CHAPTER 5

Liquid Crystal Filter LCF

Leszek R. Jaroszewicz and **Zbigniew Raszewski**[*]

Military University of Technology, Warsaw, Poland

Abstract: Multicomponent Nematic Liquid Crystalline mixtures (MNLC) containing isothiocyanato tolane and isothiocyanato terphenyl liquid crystals have been developed at the MUT. Some of them exhibit both; high optical ($\Delta n \leq 0,45$) and high dielectric ($\Delta \varepsilon \leq 20$) anisotropies and are characterized by relatively low viscosity γ. Appling the mentioned above MNLC mixtures (W1791, See Chapter 2) in HG (HomoGeneously aligned) cells with thicknesses d about 1 µm, 3 µm and 5µm, one can obtain the possibilities to develop first-order electrically tunable liquid crystal filter and three-stage ETLCF (Electrically Tunable Liquid Crystal Filter). Specific spectral filters can find some acceptance in astronomy and remote sensing for the pollution monitoring. Distinct advantages of ETLCF over conventional tunable filters are the possibility of construction of LCFs with an extremely large clear aperture, low power consumption, and low addressing voltage.

• Due to the relatively high and electrically controlled optical anisotropies $\Delta n(U)$ of MNLC and by variation of cell gaps d (1 µm, 3 µm and 5 µm) used, the LCF can select the desired wavelength $\lambda(U)$ from VIS and NIR ranges.

• Due to the high dielectric anisotropy $\Delta \varepsilon$, low viscosity γ and small cell gap d of HG cells, the LCF can achieve the response time lower than 1 ms.

In this chapter, we describe our efforts in obtaining optimization of the LCFs.

Keywords: Aperture, Dielectric anisotropy, Effective optical anisotropy, Homogeneous alignment, Liquid crystal cell, Liquid crystal filter, Nematic liquid crystalline mixture, Optical anisotropy, Ordinary refractive index, Rotational viscosity, Splay elastic constant, Switching on time.

1. INTRODUCTION

Electrically Tunable First Order Liquid Crystal Filter (LCF) was developed as an

---

[*] **Corresponding author Zbigniew Raszewski:** Military University of Technology, Faculty of Advanced Technologies and Chemistry, Warsaw, Poland; E-mail: zbigniew.raszewski@wat.edu.pl

element of air pollution detecting unit [1]. LCF, as well as three other liquid crystal indicators (Liquid Crystal Displays LCD 80E, LCD 107E, and LCD 107P), elaborated by authors of this monograph, were applied in a system of monitoring and visualization of air pollutions level. The system was installed by the Regional Inspectorate of Environment Protection (RIEP) in 1996 on the wall of the SMYK department store, Warsaw, Poland. It consisted of a transmittance measurement line running over Jerozolimskie Ave. with a light source and a detection system placed on the roofs of two buildings situated on opposite sides of the street. The LCF was installed at the front of the detection system. Due to a long distance of light (over 100 m) traveling through polluted air, the detection system should be of the aperture Φ not smaller than 160 mm and of the transmittance T not lower than 15%. Working LCF should enable an electrically driven selection of a transmitted waveband, which should not be broader than $\Delta\lambda_{\frac{1}{2}}$ = 0.1 μm within the spectrum range of $\lambda \in$ [0.5 μm - 0.7 μm]. Its switching-on time τ_{on} (time needed for selection of the given band) should not be longer than 1 ms.

A large-size display array (1700×2500 mm) with liquid crystal display LCD 80E, LCD 107E and LCD 107P working as analog indicators was fed with data indicating concentrations of: sulfur dioxide, nitrogen monoxide, ozone, toluene and formaldehyde in the air which Warsaw has been breathing. Indications have been updated every 30 s. Additionally, all indications have been averaged over a whole day and presented about acceptable standards.

2. LCF THEORY AND CONCEPT

Let us consider a transparent NLC slab of the thickness d (which simultaneously can be considered as a birefringent plate) placed in a way presented in Fig. (**1**). Let us assume that an optical axis of considered birefringent plate is collinear with the director n the NLC slab and simultaneously parallel to both boundaries surfaces S_1 and S_2 (comprising transparent electrodes of ITO). The transmittance of both a polarizer P and an analyzer A is denoted by τ [2]. Let us assume that the director n is parallel to the Ox axis of the system of coordinates. Subsequently, let us choose NLC which is a medium characterized by both positive optical anisotropy $\Delta n > 0$ and positive permittivity anisotropy $\Delta\varepsilon > 0$. Hence, the indicatrix, as well as the ellipsoid of revolution representing a static permittivity tensor, both manifest long axes n_e and $\varepsilon_{\parallel}$ to be parallel to the director n, respectively.

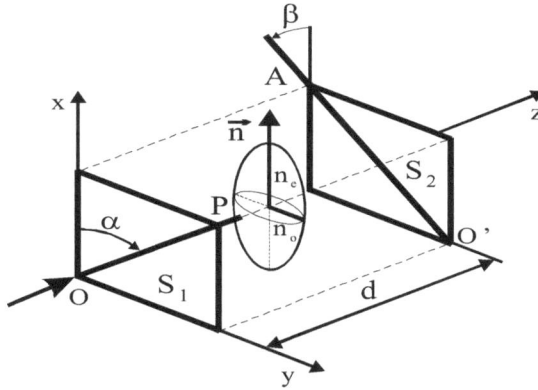

Fig. (1). Plane-parallel NLC slab of thickness d confined between a two transparent ITO electrodes S1 and S2. I_o – the intensity of unpolarized light incident the NLC layer.

According to above assumption, one can derive [3 - 5] that the incident ray of light of intensity I_0 passing through a polarizer, a birefringent NLC layer (of the optical anisotropy $\Delta n = n_e$-n_o and thickness d) and finally the analyzer, decreases to the intensity $I(\tau, \alpha, \beta, \Delta n, d, \lambda)$ defined according to the formula:

$$I = 2I_o\tau^2\left[\cos^2(\alpha - \beta) - \sin 2\alpha \sin 2\beta \sin^2\frac{\pi d\Delta n}{\lambda}\right], \qquad (1)$$

where λ is a given light wavelength and α, and β are angles defined in Fig. (1). OP and O'A determine the axes of polarizer and analyzer, respectively. The E vector (collinear with the OP segment) forms an angle β with the Ox axis. The O'A segment forms an angle α with the Ox axis. n_e denotes the extraordinary refractive index and n_o the ordinary one of the NLC slab. n_e and n_o are long and short axes of the NLC's indicatrix, respectively.

Eq. (1) will take a form of:

$$I = 2I_o\tau^2 \sin^2\frac{\pi d\Delta n}{\lambda} = I_{max}\sin^2\frac{\pi d\Delta n}{\lambda}, \qquad (2)$$

for the birefringent NLC slab with the director n (and the optical axis simultaneously) parallel to the axis Ox and between crossed polarizers oriented at the angles $\alpha = +45°$ and $\beta = -45°$. Here $I_{max} = 2I_0\tau^2$ determines the maximum intensity of light transmitted through both polarizers oriented in parallel without NLC slab between them. Hence the transmission T given by Eq. (2) is:

$$T = \frac{I}{I_{max}} = \sin^2 \frac{\pi d \cdot \Delta n}{\lambda} \qquad (3)$$

The case described above can be realized by the fabrication of a cell 1HG (with a small cell gap of $d = 1$ μm and with the homogeneous HG alignment of the director $\mathbf{n}$ on both boundary surfaces) with MNLC mixture slab characterized by a relatively high optical anisotropy $\Delta n = 0.4$. The results of measurements of the transmission spectra $T_1(\lambda)$ calculated on the basis of Eq. (3) at the value of $d\Delta n = 0.4$ μm are presented in Fig. (2). They reveal that the spectrum line of the first interference order (at $k = 1$ and the interference maximum at $\lambda = 0.8$ μm) is broad. The line width at half of the maximum is $\Delta\lambda_{1/2} = 1.0$ μm (at $d\Delta n/\lambda = \frac{1}{2}$).

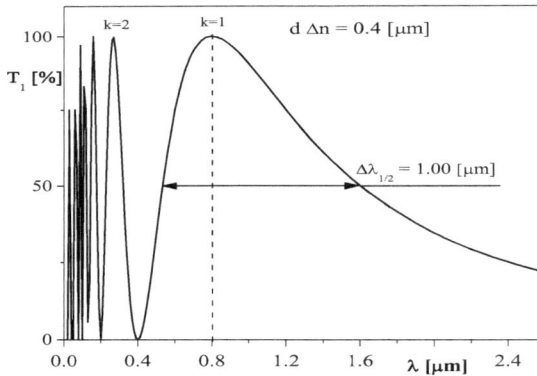

Fig. (2). The transmission spectrum $T_1(\lambda)$ of 1HG cell at $d\Delta n = 0.4$ μm.

When the factor $d\Delta n$ decreases, which takes place after application of voltage U_1 to the HG cell filled with MNLC mixture, the maximum of the first-order interference line ($k = 1$) shifts towards shorter wavelengths. When the factor $d\Delta n$ equals 0.2 μm, the first-order interference peak ($k = 1$), with its maximum at $\lambda = 0.4$ μm, is about twice as narrow as the same line at $\lambda = 0.8$ μm at $d\Delta n = 0.4$ μm (compare Figs. 2 and 3). The above example of the shift of the transmission band due to the electric field action reveals that in this way, it is possible to select a transmission band (of the first order interference) at the visible range from c.a. 0.4 μm to c.a 0.8 μm by application of proper voltage U.

In such a case, a thin 1HG cell ($d = 1$ µm) with MNLC mixture of a relatively high optical anisotropy $\Delta n = 0.4$ placed between crossed polarizers can play the role of an electrically controlled liquid crystal filter LCF operating under the first interference maximum.

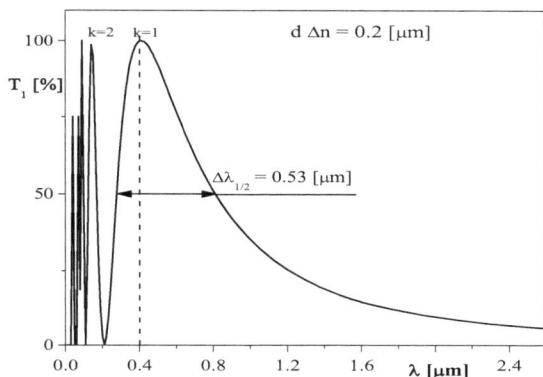

Fig. (3). The transmission spectrum $T_1(\lambda)$ of 1HG cell at $d\Delta n = 0.2$ µm.

In Fig. (**4**), the results of calculations of the transmission spectrum $T_2(\lambda)$ obtained at the value of $d\Delta n = 1.2$ µm are presented. A similar function can be practically realized by filling a 3HG cell ($d = 3$ µm) with MNLC mixture of $\Delta n = 0.4$. As one can see in Fig. (**4**), the second-order interference line ($k = 2$, $d\Delta n/\lambda = 3/2$ achieves its maximum at $\lambda = 0.8$ µm. When the factor $d\Delta n$ decreases (what can be done by applying the voltage U_2 to a 3HG cell filled with MNLC mixture of $\Delta n = 0.4$), the maximum of the second-order interference ($k = 2$) shifts towards shorter wavelengths. At $d\Delta n = 0.6$ µm, the second-order interference line ($k = 2$) with its maximum at $\lambda = 0.4$ µm is about twice as narrow as the same line at $\lambda = 0.8$ µm for $d\Delta n = 1.2$ µm (compare Figs. **4** and **5**).

In Fig. (**6**), the results of calculations of the transmission $T_3(\lambda)$ for the value of $d\Delta n = 2.2$ µm are presented. Such a case can be practically realized by filling the 5HG cell ($d = 5$ µm) with MNLC mixture of $\Delta n = 0.4$. As one can see in Fig. (**6**), the third-order interference line present for ($k = 3$, $d\Delta n/\lambda = 5/2$) attains its maximum at $\lambda = 0.8$ µm. Again, if the factor $d\Delta n$ decreases (what can be practically realized by applying the voltage U_3 to the 3HG cell filled with MNLC mixture of $\Delta n = 0.4$), the maximum of the third-order interference ($k = 3$) shifts towards shorter wavelengths. When $d\Delta n = 2.2$, the third-order interference line ($k = 3$) with its maximum at $\lambda = 0.4$ µm is about twice as narrow as the same line at $\lambda = 0.8$ µm at $d\Delta n = 0.4$ µm (see Figs. **6** and **7**).

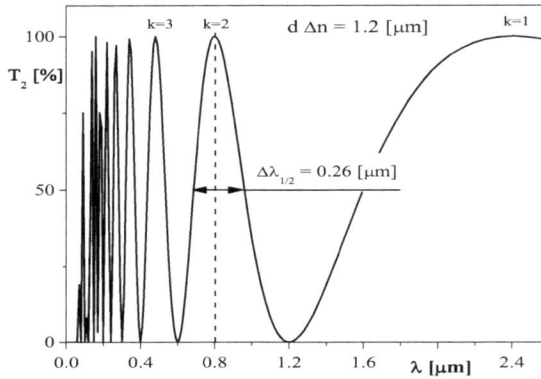

Fig. (4). The transmission spectrum $T_2(\lambda)$ of 3HG cell at $d\Delta n = 1.2$ µm.

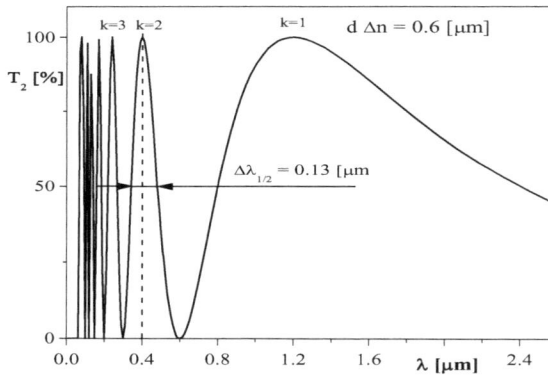

Fig. (5). The transmission spectrum $T_2(\lambda)$ of 3HG cell at $d\Delta n = 0.6$ µm.

The transmission spectra $T_2(\lambda)$ for 3HG cell filled with MNLC mixture of $\Delta n = 0.4$ presented in Figs. (**4** and **5**) reveal that this cell affected with the voltage U_2 shifts the second-order line maximum ($k = 2$) within the range of λ from 0.4 µm to 0.8 µm, what is accompanied by a shift of two neighboring spectral lines $k = 1$ and $k = 3$. The same phenomenon can be observed in Figs. (**6** and **7**). Here for the 3HG cell driven by the voltage U_3, a shift of the third order line ($k = 3$) within the visible range goes together with the appearance of the line with $k = 2$, $k = 4$ and $k = 5$.

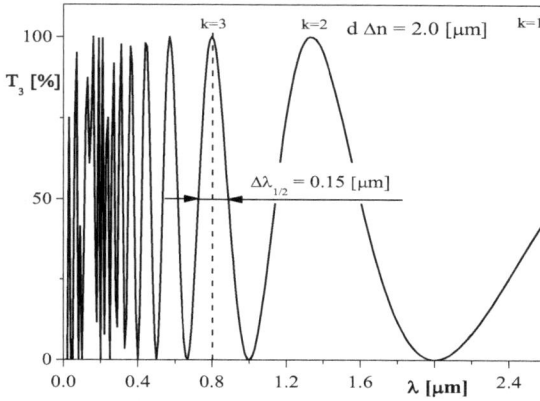

Fig. (6). The transmission spectrum $T_3(\lambda)$ of 5HG cell at $d\Delta n = 2.0$ μm.

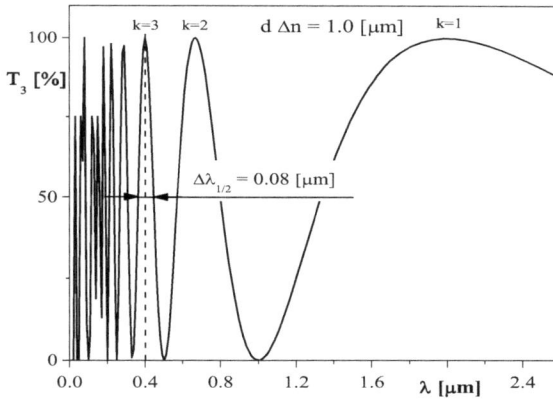

Fig. (7). The transmission spectrum $T_3(\lambda)$ of 5HG cell at $d\Delta n = 1.0$ μm.

Taking into consideration the above, a structure of electrically controlled liquid crystal filter LCF can be proposed. Schematic diagram of the LCF is presented in Fig. (**8**). The LCF consists of three crystal cells (1HG, 3HG, and 5HG) filled with MNLC mixture of $\Delta n = 0.4$ controlled by voltages U_1, U_2, U_3, respectively. Those cells are placed between crossed polarizers so that each of them is placed in a system of crossed polarizer and analyzer.

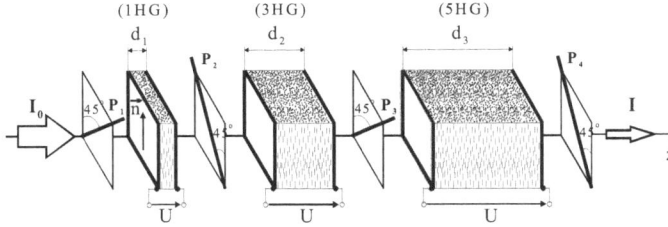

Fig. (8). Schematic diagram of LCF consisting of three liquid crystal cells (1HG, 3HG and 5HG) filled with NLCM, placed between four crossed polarizers (P_1, P_2, P_3 and P_4) and controlled with voltages U_1, U_2 and U_3.

The light intensity I passing through seven-layer LCF consisting of three liquid crystal cells (1HG, 3HG, and 5HG) placed between four crossed polarizers (P_1, P_2, P_3, and P_4) can be written as:

$$I = 2\left\{2\left[2I_o\tau \ \sin^2\left(\frac{\pi d_1\Delta n_{1ef}}{\lambda}\right)\right]\tau \sin^2\left(\frac{\pi d_2\Delta n_{2ef}}{\lambda}\right)\right\}\tau \sin^2\left(\frac{\pi d_3\Delta n_{3ef}}{\lambda}\right)\tau \qquad (4)$$

where Δn_{ief} is the effective optical anisotropy of i-*th* HG cell (cell gap d_i) driven with voltage U_i (i = 1, 2, 3).

At the first approximation, the effective optical anisotropy Δn_{ief} of an HG cell, of a cell gap d_i, with MNLC mixture of $\Delta n > 0$ and $\Delta\varepsilon > 0$ driven with a voltage U_i (see Fig. **9**) can be written as [6]:

$$\Delta n_{ief} = \frac{n_o n_e}{\sqrt{n_o^2 \cos^2 \Theta_i + n_e^2 \sin^2 \Theta_i}} - n_o , \qquad (5)$$

where Θ_i is the angle between the director $\boldsymbol{n}$ and the Ox axis averaged over the entire cell gap d_i.

According to Chapter 3, the MNLC mixture is characterized by the following material parameters: K_{11} –splay deformation elastic constant, γ – rotational viscosity, n_e – extraordinary and n_o – ordinary light refractive indices, $\varepsilon_\perp$ – perpendicular component and $\varepsilon_\parallel$ – parallel component of permittivity tensor. In the cell gap filled with MNLC mixture of $\Delta\varepsilon > 0$, bounded by ITO electrodes, the $\boldsymbol{n}_i(z)$ director field is formed after application of voltage U_i [7, 8]. This $\boldsymbol{n}_i(z)$ field, at the first approximation, can be described by the constant angle Θ_i averaged over the entire thickness d_i.

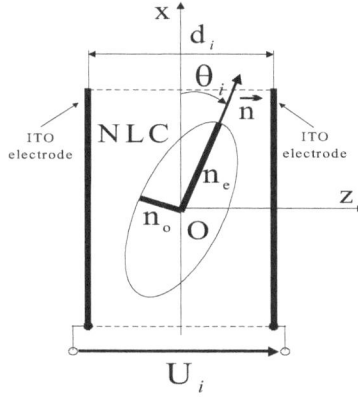

Fig. (9). The HG cell with MNLC mixture of $\Delta n > 0$ and $\Delta\varepsilon > 0$ driven with voltage U_i. The cell gap is d_i.

One can notice (see Eq. (4)) that LCF at no voltage applied, all three MNLC mixture layers show the same values of effective optical anisotropy: $\Delta n_{1\mathrm{ef}} = \Delta n_{2\mathrm{ef}} = \Delta n_{3\mathrm{ef}} = \Delta n = 0.4$. Here the light transmission T_{FO} through LCF can be calculated as:

$$T_{FO} = T_1 \cdot T_2 \cdot T_3 = \frac{I}{I_{F\max}} \tag{6}$$

with

$$T_i = \sin^2 \frac{\pi d_i \Delta n}{\lambda}, \quad i = 1, 2, 3, \tag{7a}$$

$$I_{F\max} = 8 I_o \tau^4. \tag{7b}$$

Fig. (**10**) displays the spectrum of the total transmission $T_{\mathrm{FO}}(\lambda)$ calculated based on Eq. (6) for LCF consisting of three cells of cell gaps $d_1 = 1$ μm, $d_2 = 3$ μm, $d_3 = 5$ μm, respectively at no voltage applied; $U_1 = U_2 = U_3 = 0$; in other words at a state at the absence of control voltages. Here all MNLC mixture layers are undeformed and are of the same effective optical anisotropy $\Delta n_1 = \Delta n_2 = \Delta n_3 = 0.4$.

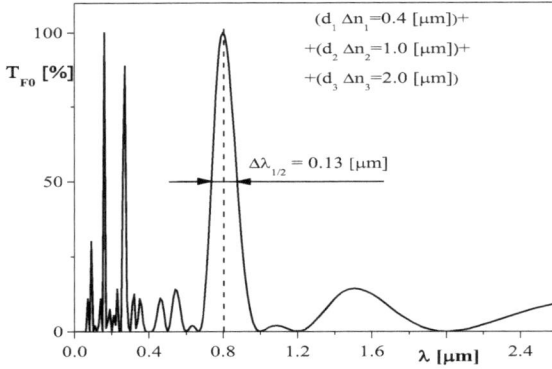

Fig. (10). The transmission spectrum $T_{FU}(\lambda)$ of the LCF at $U_1 = U_2 = U_3 = 0$.

When LCF's cells are driven with chosen controlling voltages, namely: U_1 applied to the cell 1HG, in order to obtain $\Delta n_{1\text{ef}}(U_1) \cdot d = 0.2$ µm; U_2 applied to the cell 3HG, in order to obtain $\Delta n_{2\text{ef}}(U_2) \cdot d = 0.6$ µm; U_3 applied to the cell 5HG, in order to obtain $\Delta n_{3\text{ef}}(U_1) \cdot d = 1.0$ µm, then LCF forms the transmission spectrum T_{FU} presented in Fig. (**11**).

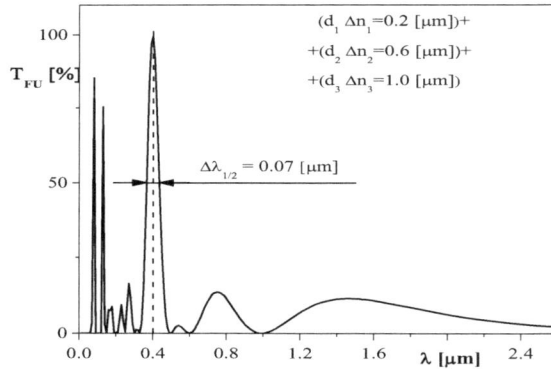

Fig. (11). The transmission spectrum $T_{FU}(\lambda)$ of LCF at U_1, U_2, and U_3 according to relation shown in a top right corner of the picture.

The above analysis describes phenomena occurring in the driven, three-cell LCF. It proves that by simultaneous application of the selected voltages U_i to proper cells, one can select transmission at a chosen wave band with a maximum of transmission at $\lambda(U_i)$ located not only at visible but also at the near-infrared range.

One should notice that the transmission's band half-width is $\Delta\lambda_{1/2} = 0.13$ μm for the maximum position at $\lambda = 0.8$μm (voltage free cells) and under the action of the driving electric field, it is gradually narrowing up to value $\Delta\lambda_{1/2} = 0.07$ μm for $\lambda = 0.4$ μm. One can assume that LCF described above will meet the set of technical requirements of air pollution diagnostics unit and will provide an analog selection of light band not wider that $\Delta\lambda_{1/2} \approx 0.1$ μm within a spectrum range of $\lambda \in [0.5$ μm, 0.7 μm$]$.

Let us assume the following material parameters of the MNLC mixture for LCF: $n_0 \sim 1.5$, $n_e \sim 1.9$, $K_{11} \sim 20 \cdot 10^{-12}$ N, $\gamma \sim 150$ mPa·s, $\varepsilon_\perp \sim 4.0$, $\varepsilon_\| \sim 19.0$. A switching-on time τ_{on} for selecting by LCF a given wave band with its maximum at λ would be determined by the total switching time (time needed for the director angle to shift Θ from $0°$ to $90°$) in an HG cell ($d = 5$ μm) under nominal voltage $U = 5$ V_{RMS} and period $T = 10$ ms (see Figs. **9** and **12**). The switching-on time $\tau_{on} = 1$ ms marked in Fig. (**12**) was evaluated on the basis of the following relation:

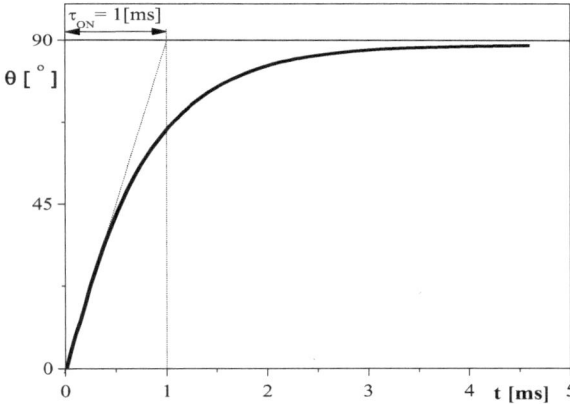

Fig. (12). Switching process of angle Θ at a 5HG cell ($d = 5$ μm) filled with MNLC mixture with the following material parameters: $K_{11} \sim 20 \cdot 10^{-12}$ N, $\gamma \sim 150$ mPa·s, $\varepsilon_\perp \sim 4.0$ and $\varepsilon_{II} \sim 19.0$ driven with a voltage of $U = 5$ V.

$$\tau_{ON} \propto \frac{\gamma d^2}{\Delta\varepsilon\varepsilon_o U^2}. \qquad (8)$$

It results from Eq. (5) that to make a significant change of Δn_{ef} from 0.4 to 0.2 in the thickest liquid crystal cell 5HG, the angle Θ has to increase from $0°$ to $40°$ under the voltage of $U = 5$ V_{RMS}. In such a case, one can be sure that switching-on time τ_{ON} will not be longer than 1 ms, so it meets technical requirements for response time set of LCF dedicated to air pollution diagnostics system.

Taking into consideration Eq. (7b) and assuming that all four polarizers of the LCF are characterized by the same transmission coefficient $\tau = 0.4$ [2], one can evaluate the maximum intensities I_{Fmax} of light transmitted by LCF as $I_{Fmax} = 8I_0\tau^4 = 0.2I_0$. One can see that intensity of light I at a wavelength λ passing through the LCF represents only 20% of the incident light intensity I_0. Nevertheless, it meets technical requirements for the minimum transmission $T > 15\%$ of a spectral filter for the air pollution diagnostics unit.

3. LIQUID CRYSTAL FILTER LCF – AN WORKING EXAMPLE

From the theoretical considerations and numerical simulations, a set of material and technical parameters was evaluated. For the fabrication of a test example of the LCF it was decided to:

- design, synthesize compounds and compose a new, tailored MNLCF mixture of material parameters like: $n_o \sim 1.5$ at $\lambda = 0.8$ µm, $\Delta n > 0.4$ at $\lambda = 0.8$ µm, $\varepsilon_\perp \sim 4.0$ at $f = 1.5$ kHz, $\Delta\varepsilon > 15.0$ at $f = 1.5$ kHz, $K_{11} \sim 20 \cdot 10^{-12}$ N, and rotational viscosity $\gamma \sim 150$ mPa·s, where the viscosity γ has been estimated as the lowest one possible to achieve without a significant decrease of $\Delta\varepsilon$,
- manufacture a three-cell LCF of the aperture Φ not lower than 160 mm according to the schematic diagram presented in Fig. (8) to meet all the technical requirements set by RIEP for liquid crystal filter dedicated for air pollution diagnostics system.

3.1. Multicomponent Nematic Liquid Crystalline Mixture for LCF

According to the assumptions formulated above, a team of chemists from the Institute of Chemistry of the MUT composed the mixture abbreviated W1791, consisting of fluorine-isothiocyanates, alkyltolanes and mainly alkylophenyl-tolanes. It exhibited the phase sequence of **Cr**- 20.0°C -**N**- 127.5°C -**Iso**. The W1791 mixture was extensively studied in the Institute of Applied Physics of the MUT [9].

Detailed results of the mixture W1791 study have been presented in Chapter 3. Selected material parameters of this mixture for LCF are gathered in Table 1.

Table 1. Material parameters of the liquid crystal mixture W1791 for LCF evaluated at 25°C.

Parameter	Value, Unit
Optical anisotropy Δn (at $\lambda = 0.45$ µm)	0.50
Optical anisotropy Δn (at $\lambda = 0.80$ µm)	0.42
Ordinary refractive index n_o (at $\lambda = 0.589$ µm)	1.54
Dielectric anisotropy $\Delta\varepsilon$ (at $f = 1.5$ Hz)	16.4
Perpendicular component of permittivity tensor - $\varepsilon_\perp$	4.5
Splay elastic deformation constant K_{11}	21.2 pN
Rotational viscosity γ	155 mPa·s

The Table above reveals that material parameters of the mixture W1791 meet the requirements for the working, tailored MNLC mixture of LCF for air pollution diagnostics system.

3.2. Liquid Crystal Filter – Test's Results

It was decided to manufacture three HG cells with cell gaps slightly braoder than those proposed in Section "LCF theory and concept". That has been done due to the availability of the mixture W1791 of optical anisotropy of $\Delta n \sim 0.42$, *i.e.* a little more than this assumed at the theoretical analysis ($\Delta n \sim 0.40$), necessary for LCF tuning at wavelengths $\lambda \in [0.5$ µm, 0.7 µm]. Moreover, a big aperture of LCF ($\Phi = 160$ mm) resulted in high demands for the technological regime concerned at the cell gap as small as $d=1$ µm. Moreover, such cells need to be equipped with insulating layers made of SiO_2. Finally, HG cells (of a structure presented in Fig. (**13**)) of cell gaps of $d = 1.8$ µm, $d = 3.5$ µm and $d = 5.0$ µm, still tagged as 1.8HG, 3.5HG and 5HG were fabricated. Those cells (before being filled with MNLCF) exhibited a total resistance R not less than 3 MΩ. After filling them with MNLCF of the total specific resistivity of $\rho \sim 1 \cdot 10^8$ Ωm, the total resistance R of fabricated cells at a frequency of $f = 1.5$ kHz of the driving electric field was not lower than 1 MΩ. That enabled the induction of Electrically Controlled Birefringence effect (ECB) or forcing (by electric voltage U) a shift of the effective optical anisotropy $\Delta n(U)$, needed for tuning the wavelength λ in LCF. Substrates of all HG cells used for assembling of LCF were made of float glass (by Balzers) of a thickness of 1.1 mm. The active surface of the substrate was covered with a 20 nm of SiO_2 blocking layer and c.a. 80 nm ITO electrode layer of specific resistance of 100 Ω/□. To obtain HG alignment, the ITO electrodes, covered with SiO_2 insulating layers, were spin-coated with polyimide alignment layers of PI 2610 (by du Pont) polyimide and next, they were dried and cured at elevated temperature for polymerization. Subsequently, they were unidirectionally rubbed.

In order to obtain the desired cell gap d of HG cells, the glass spacers of diameters d - 1.8, 3.5 and 5.0 μm were added to the adhesive gasket path and the same spacers were spread over a whole surface of substrates in the amount of 2 - 4 pcs/mm^2. Such substrates were assembled in the form of a flat-parallel cell which was formed under a hot press. The final cell gap of individual HG cells was determined by the interference method. After filling and sealing of all the cells; 1.8HG, 3.5HG and 5HG, they were stacked together with relevant polarizers (NPF-F1225DUAG20 by NITTO, $\tau = 44.5\%$) as presented in Fig. (**13**). Two external surfaces of used cells were armed with anti-reflective layers, and two middle ones had no such layer.

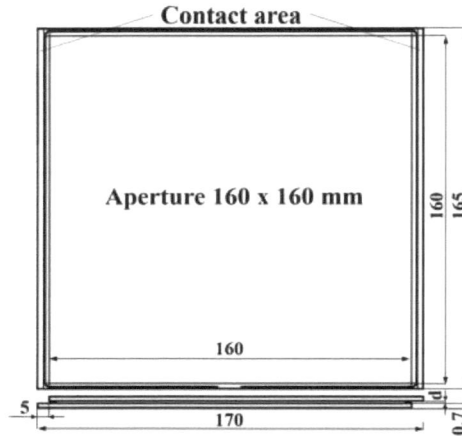

Fig. (13). Layout of the HG cells for LCF.

An effective optical anisotropy $\Delta n_{\text{eff}}(U)$ of the nematic W1791 slab in 1.8HG, 3.5HG, and 5HG cells was achieved by application of an appropriate electric voltages U in the form of rectangular waves of amplitudes U_1, U_2 and U_3 at the frequency $f = 50$ Hz. Each cell was controlled by a separate pulse Generator HP33120A.

$T_1(\lambda)$, $T_2(\lambda)$ and $T_3(\lambda)$, the transmission spectra obtained using a BRC111A spectrometer of a single liquid crystal cells 1.8HG, 3.5HG and 5HG with at voltages $U_1 = 2.41$ V, $U_2 = 1.95$ V and $U_3 = 1.65$ V respectively, are presented in Figs. (**14 - 16**). In Figs. (**17 - 19**), three transmission $T_F(\lambda)$ spectra obtained by using BRC111A spectrometer, tuned by three different sets of voltages U_1, U_2 and U_3, at which LCF selected bands at $\lambda_1 = 0.7$ μm, $\lambda_2 = 0.6$ μm and $\lambda_3 = 0.5$ μm are presented. All measurements of transmissions $T_1(\lambda)$, $T_2(\lambda)$ and $T_3(\lambda)$ of each separate cell, as well as TF(λ) of LCF filter, were carried out at room temperature. At the voltage, $U_1 = 2.41$ V applied to the 1.8HG cell with W1791 in the spectrum $T_1(\lambda)$ of that cell, one can observe a band of the first interference maximum ($k = 1$) with its maximum transmission at $\lambda = 0.7$ µm (see Fig. **14**).

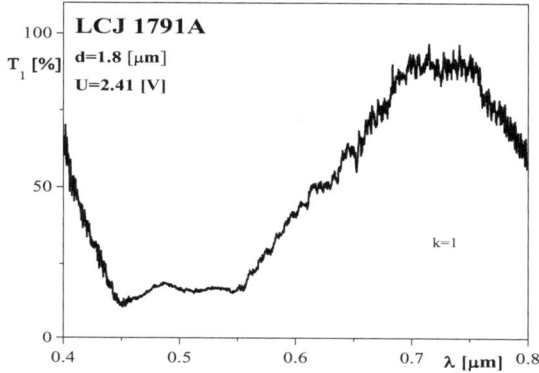

Fig. (14). The transmission spectrum $T_1(\lambda)$ of 1.8HG cell at $U_1 = 2.41$ V.

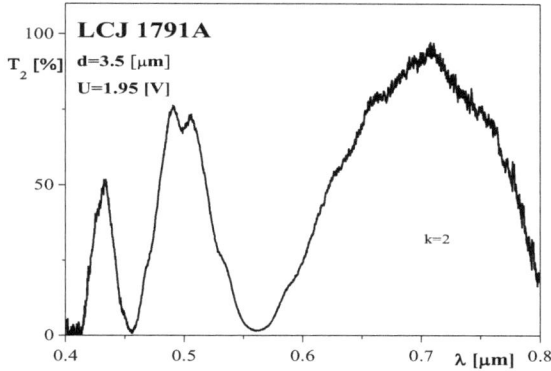

Fig. (15). The transmission spectrum $T_2(\lambda)$ of 3.5HG at for $U_2 = 1.95$ V.

In Figs. (**15** and **16**), one can see that at the voltage of $U_2 = 1.95$ V applied to the cells 3.5HG and voltage of $U_3 = 1.65$ V applied to cell 5HG, bands of the second ($k = 2$) and third ($k = 3$) interference maximum in the spectra $T_2(\lambda)$ and $T_3(\lambda)$ of these cells respectively overlap at the same wavelength of $\lambda = 0.7$ µm as in the case of 1.8HG cell. Hence, if the white light enters the stack of cells 1.8HG, 3.5HG and 5HG with MNLCF attuned by the voltages $U_1 = 2.41$ V, $U_2 = 1.95$ V and $U_3 = 1.65$ V respectively, one can observe the transmission of light at a single band at $\lambda = 0.70$ µm what can be seen in Fig. (**17**). According to Eq. (6), one can tune the transmission band $T_F(\lambda)$ at the entire visible range by the selection of voltage settings U_1, U_2 and U_3 (see Table **2**). Since voltage control functions $\lambda = f(U_1)$, $\lambda = f(U_2)$, $\lambda = f(U_3)$ for three cells of LCF have nearly linear character, then controlling of the transmission band is straightforward.

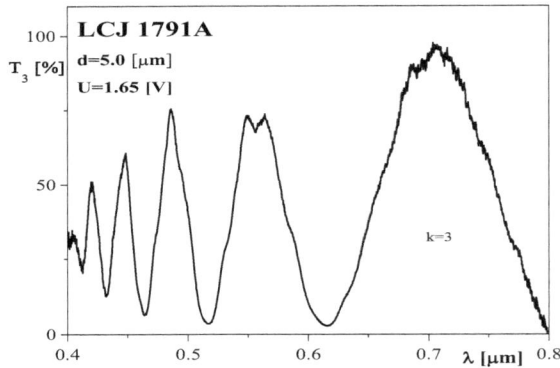

Fig. (16). The transmission spectrum $T_3(\lambda)$ of 5HG cell at $U_3 = 1.65$ V.

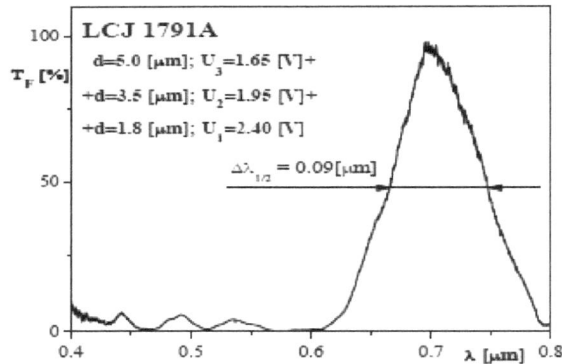

Fig. (17). The transmission spectrum $T_F(\lambda)$ of the LCF filter at $U_1 = 2.41$ V, $U_2 = 1.95$ V and $U_3 = 1.65$ V.

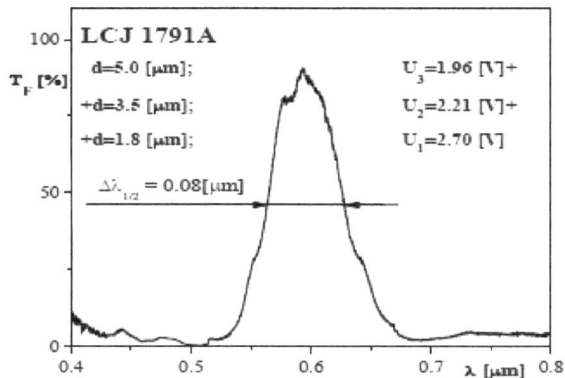

Fig. (18). The transmission spectrum $T_F(\lambda)$ of the LCF filter at $U_1 = 2.71$ V, $U_2 = 2.20$ V and $U_3 = 1.96$ V.

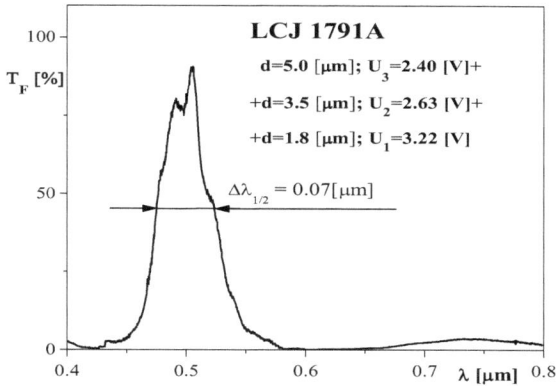

Fig. (19). The transmission spectrum $T_F(\lambda)$ of the LCF filter at $U_1 = 3.03$ V, $U_2 = 2.62$ V and $U_3 = 2.31$ V.

Table 2. Examples of driving voltages U_1, U_2 and U_3 controlling liquid crystal LCF with MNLCF at 22.5 °C.

	$\lambda = 0.70$ μm	$\lambda = 0.65$ μm	$\lambda = 0.60$ μm	$\lambda = 0.55$ μm	$\lambda = 0.50$ μm
U_1 [V]	2.41	2.57	2.71	2.85	3.03
U_2 [V]	1.95	2.09	2.20	2.42	2.62
U_3 [V]	1.65	1.80	1.96	2.21	2.31

Detailed measurements reveal that the relative transmission $T(\lambda)$ at the maximum of the transmission band of the LCF is surely higher than 16%, which meets the technical requirements of LCF for air pollution diagnostics.

CONCLUSION

Despite the fact that LCF described in this chapter selects rather broad bands ($\Delta\lambda_{1/2} \sim 0.1$ μm), it was successfully applied in remote sensing and air pollution monitoring system. The system operated flawlessly in the public area for three years, and, then was uninstalled while the change of SMYK's ownership. Nevertheless, LCF selects a rather broad waveband. The LCF described above offers the extremely high optical apertures Φ available, low values and linearity of controlling voltages as well as low power consumption, which are undoubtedly valuable advantages over other commercially available automatic visible light filters with extremely high apertures. There are theoretical and technical possibilities of creating an LCF consisting of four or even five HG cells. If one adds to the structure described above the fourth 7HG cell, full width at half maximum of a filtered band will decrease to $\Delta\lambda_{1/2} \sim 0.04$ μm, but the transmitted light intensity will decline to value as low as 6%.

CONSENT FOR PUBLICATION

Not applicable.

CONFLICT OF INTEREST

The author(s) confirms that there is no conflict of interest.

ACKNOWLEDGEMENTS

Declared none.

REFERENCES

[1]   Z. Raszewski, E. Nowinowski, J. Kędzierski, P. Perkowski, W. Piecek, R. Dąbrowski, P. Morawiak, and K. Ogrodnik, "Electrically Tunable Liquid Crystal Filters", *Mol. Cryst. Liq. Cryst. (Phila. Pa.)*, vol. 525, pp. 112-127, 2010.
[http://dx.doi.org/10.1080/15421401003796132]

[2]   D. Clarke, and J.F. Grainger, *Polarized Light and Optical Measurements*. Pergamon Press: Oxford, 1971.

[3]   M. Born, and E. Wolf, "*Principles of optics*", 17th Edition - Electromagnetic Theory of Propagation, Interference and Diffraction of Light. London: Cambridge University Press, 2003.

[4]   M. Pluta, Advanced Light Microscopy: Measuring Techniques *Vol. 3*, Elsevier Science: Amsterdam, 1993.

[5]   M. Pluta, "Simplified polanret system for microscopy", *Appl. Opt.,* vol. 28, no. 8, pp. 1453-1466, 1989.
[http://dx.doi.org/10.1364/AO.28.001453] [PMID: 20548681]

[6]   E. Miszczyk, Z. Raszewski, J. Kędzierski, E. Nowinowski-Kruszelnicki, M.A. Kojdecki, P. Perkowski, W. Piecek, and M. Olifierczuk, "Interference method for determination of refractive indices of liquid crystal", *Mol. Cryst. Liq. Cryst. (Phila. Pa.),* vol. 544, pp. 22-36, 2011.
[http://dx.doi.org/10.1080/15421406.2011.569262]

[7]   J. Kędzierski, M.A. Kojdecki, Z. Raszewski, J. Zieliński, and L. Lipińska, "Determination of anchoring energy, diamagnetic susceptibility anisotropy, and elasticity of some nematics by means of semiempirical method of self-consistent director field", *Proc. SPIE,* vol. 6023, pp. 26-40, 2005.
[http://dx.doi.org/10.1117/12.648167]

[8]   J. Kędzierski, M.A. Kojdecki, Z. Raszewski, J. Rutkowska, W. Piecek, P. Perkowski, J. Zieliński, and E. Miszczyk, "Study of anchoring characteristics and splay-bend elastic constant based on experiments with non-twisted nematic liquid crystal cells", *Opto-Electron. Rev.,* vol. 16, no. 4, pp. 390-394, 2008.
[http://dx.doi.org/10.2478/s11772-008-0046-8]

[9]   E. Nowinowski-Kruszelnicki, J. Kędzierski, Z. Raszewski, L. Jaroszewicz, R. Dąbrowski, W. Piecek, P. Perkowski, M. Olifierczuk, K. Garbat, M. Sutkowski, E. Miszczyk, K. Ogrodnik, P. Morawiak, M. Laska, and R. Mazur, "High birefringence liquid crystal mixtures for lc electro-optical devices", *Opt. Appl.,* vol. 42, no. 1, pp. 167-180, 2012.

CHAPTER 6

Liquid Crystal Shutter for Welding Helmets "PIAP-PS Automatic"

Paweł Perkowski[*]

Military University of Technology, Warsaw, Poland

Abstract: In this chapter, our efforts to obtain and optimize Liquid Crystal Shutter (LCS) for switching a light transmission in welding helmets have been described and discussed. Automatic protection devices in the welding helmet should have sufficiently short switching time and extremely low transmission, preventing the absorption of harmful doses of radiation regardless of light direction and its polarization. A short time of return to the initial off-state is required as well. LCS operating in welding helmets has a form of a removable module consisting of a stack of two liquid crystal cells and an electronic control system. In normal illumination conditions, at off-state, this module is transparent, which allows observation of welded objects. In the presence of an extremely intensive illumination, caused by an electric arc, the module rapidly reduces the transparency, with the intensity of the light reaching welder's eyes. Application of automatic helmets increases welder's safety and, by making both hands free, also increases their performance. According to the idea formulated above, we developed the LCS. The LCS consisted of two liquid crystal cells (of cell gap d = 6 μm) filled with a W1115 mixture ($\Delta n = 0.08$ at $\lambda = 0.589$ μm, see Chapter 3) placed between three crossed polarizers and a band-pass Dielectric-Metallic Interference Filter (DMIF). Our LCS fully meets the requirements to be applied in professional welding helmets with protection degree up to 13 N.

Keywords: Aperture, Contrast ratio, Degree of protection, Dielectric anisotropy, Driving voltage, Liquid crystal cell, Liquid crystal shutter, Optical anisotropy, Ordinary refractive index, Reduced elastic constant, Rotational viscosity, Transmission, Twisted nematic effect, Switching-on time.

1. INTRODUCTION

Performing a special work is sometimes linked to a periodic or continuous exposure of sight to illumination when intensity exceeds a safe level. While considering the influence of electromagnetic radiation on a human being, it is

---
[*] **Corresponding author Paweł Perkowski:** Military University of Technology, Faculty of Advanced Technologies and Chemistry, Warsaw, Poland; E-mail: pawel.perkowski@wat.edu.pl

Leszek R. Jaroszewicz (Ed.)

assumed that the most exposed part of the human body is the face with eyes and eyelids. Automatic protection devices should have a sufficiently short response time and the transmission ratio preventing the absorption of harmful doses of radiation regardless of light direction and its polarization. A short time of returning to the initial state is required as well. Conventional solutions rely on the use of filters with constant attenuation coefficient modulated as a function of wavelength adapted to the particular application. For the user's comfort, the eye protection devices should not impede work but rather foster productivity growth. When working with lasers operating at the infrared range (IR) one uses eye protection transducers. They are image intensifiers (pretending night vision devices) which, on the one hand, protect the sight, and allow beam observation. Image transducer is a special type of a vacuum tube where an image is projected on a photocathode. Electrons emitted from photocathode are accelerated towards a luminescent screen where an image is created. The device protects the sight by constituting a physical barrier. Unfortunately, photocathodes of image intensifiers are susceptible to visual over-brightness in response to intensive pulses of laser light, causing long relaxation time, too long in many applications. At the VIS spectrum, devices using photochromic effects appear to be too slow and protection coefficients too low for the majority of applications involving dazzling. A modern solution is applying materials with non-linear characteristics of the absorption coefficient. A filter made of such a material is transparent for the light within the low-intensity area but blocks high-intensity radiation with a response time of several picoseconds [www.fujipoly.com]. Such a solution has many advantages. It requires no power and can be applied for protection in a wide spectral range of the spectrum with very short response times. Unfortunately, no parameters allowing application in eye protection systems were obtained. Welding masks with automatic liquid crystal shutters appeared on the market in the late 1970s, (US Patent 4.039.234 dated August 2nd, 1977 "Electro-optic welding lens assembly using multiple liquid crystal light shutters and polarizers"). Automatic shutters operating in welding helmets have a form of a removable module consisting of optical elements and electric control system. In normal illumination conditions, this module is transparent, which allows observation of welded objects. In the presence of an extremely intensive illumination, caused by an electric arc, the module radically reduces transparency, hence the intensity of the light reaching welder's eyes. Application of automatic helmets increases welder's safety, and by making hands free also increases work performance. Moreover, such eye protection allows welding in a confined space, where there is not enough space for moving hands freely.

The intensity of emitted radiation and its spectral characteristics depends on the applied welding technology and characteristics of welded materials. Electronically controlled liquid crystal elements allow automatic adjustment of the transmission degree, thereby, allowing a degree of eye protection regardless of the welding method.

2. TECHNICAL REQUIREMENTS FOR WELDING FILTERS

Requirements for welding filters which automatically tune their light transmission coefficient to a lower value are given by Polish standard PN-EN 379 – *"Requirements concerning welding filters with switching light transmission and welding filters with two light transmission coefficients"* as well as by European standards: PN-EN 165 - *"Personal eye protecting-Definitions"*, PN-EN 166 - *"Requirements"*, PN-EN 167 - *"Optical tests' methods"*, PN-EN 168 - *"Non-optical tests' methods"*, PN-EN 169 – *"Personal eye-protecting – welding filters and filters for related techniques"*, *"Requirements concerning light transmission coefficient and recommended use"*, PN-EN 175 - *"Personal protecting – eye and face protection while welding and in related processes"*. These standards comprise general requirements that should be met by an optical element. These standards strictly define:

- spherical power,
- astigmatic power defined as refractive power (positive or negative) occurring in the form of residual resulting from implementation inaccuracy of "the surface" of valve window element,
- a difference of prismatic power resulting from valve window mounting uncertainty,
- heterogeneity of light transmission coefficient,
- dispersion,
- stability at high temperature,
- resistance to ultraviolet.

The module comprising an optical filter and an electronic control system is mounted in a welding helmet, protecting the neck and ears of a welder as well. In advanced models of welding helmets, air filters are mounted. Helmets, before being released for use, have to get a certificate confirming meeting all the requirements mentioned above. In Poland, such certificates are issued after testing an item by the Central Institute for Labor Protection (CIOP). Liquid crystal filter is the most important and complicated part of the welding helmet. Optical parameters of the filter, switching times and total transmission coefficient in particular, to a large extent, depend on the electronic control system. In regular illumination conditions, a filter has a transmission degree corresponding to the highest value of light transmission coefficient (degree of protection in the white state). In this "open" state environment, observation is possible. Initiation of welding arc leads to a stepwise increase of light intensity which is detected by a photodetector which initiates the application of a voltage with the appropriate value and shapes filter controlling electrodes causing passing into the "close" state with the lowest value of light transmission coefficient (degree of protection in a black state).

In the process of scientific development of a liquid crystal welding helmet design, the authors mainly focused on developing and manufacturing an MNLC mixture, and electrically controlled liquid crystal valve element Liquid Crystal Shutter (LCS), supplemented with the appropriate filters [1]:

- the whole module (optical elements assembly along with the control system creating the vision protection device) completely suppress radiation in spectrum areas outside VIS range - IR (700 nm to 12000 nm) and UV (from 100 nm to 420 nm) where protection degree (in the black state) was regulated in the range of 9N to 13N,
- shade degree at an off-state was higher than 4N,
- switching time from transparent off-state to blocking on-state t_{on} was not longer than 0.2 ms,
- temperature range of proper performance was in a range from -5°C to +60°C,
- minimum dimensions of the valve area after mounting in hardware were not lower than 100x43 mm.

For commercial success, the requirement should be met however; maximum current consumption by the whole LCS switched-on should not be higher than 60 μA. This allows the use of available photovoltaic cells illuminated by a welding arc for the power supply of the helmet.

It must be noted that LCS, along with polarizers and permanent protecting filters, photodetector, module LCS controlling electronics as well as a system of photovoltaic cells, were characterized by affordability promising success in the market.

3. DEGREE N OF FILTER PROTECTION

Degree N of filter protection is a number which has (Polish standard PN-81, and in Table **3**) a specified range of values and a nominal value of the total light transmission coefficient passing through the filter. Degree N of filter protection is defined by the following equation:

$$N = \frac{7(-\log_{10}\tau)}{3} + 1, \tag{1}$$

where τ is the total transmission coefficient of the filter with the spectral characteristics $\tau(\lambda)$ measured by a detector with the relative luminous efficiency $V(\lambda)$ for optic vision, lighted with a light source with the relative value of spectral density of the luminous flux $\Phi_{e\lambda}(\lambda)$ of a standard illuminant A, expressed by the equation:

$$\tau = \frac{\int_{380nm}^{780nm} \Phi_{e\lambda}\tau(\lambda)V(\lambda)d\lambda}{\int_{380nm}^{780nm} \Phi_{e\lambda}V(\lambda)d\lambda}. \tag{2}$$

4. SWITCHING TIME T_S

Switching time (switching time into blocking state) t_s is the time between the moment where a detector detects an arc inflammation and the other moment when a sufficiently low degree of transparency of the filter is reached. The switching time t_s measurement method, described in the standard EN-379, specifies the t_s time with the following integral:

$$t_s = \frac{1}{\tau_1} \int_{t=0}^{t=3\tau_2} \tau(t)\, dt, \tag{3}$$

where $t = 0$ is the moment when illumination intensity registered on a given element reaches half the maximum value, time $t = 3\tau_2$ is the time where light transmission coefficient will decrease to a degree of a triple transmission coefficient value at the black state. The light transmission coefficients τ_1 and τ_2 represent respectively: the maximum value (at transparent off-state) and the minimum value of transmission coefficient (at blocking on-state).

5. DESIGN AND OPERATION PRINCIPLE OF THE LCS

To meet technical requirements formulated above a liquid crystal valve LCS (called filter also) was designed (see in Fig. 1). LCS consists of a stack of two liquid crystal cells (3), three (polarizers) polarizing polymer films (2), plates with thin-layer, a dielectric filter deposited on the internal surface (1) and a protective plate (4).

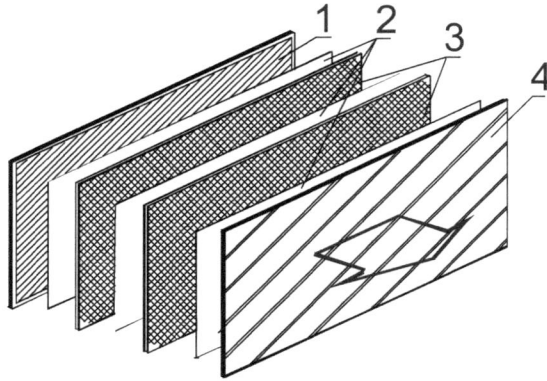

Fig. (1). Scheme of LCS for a welding helmet. 1 - plate with dielectric-metal (permanent) band filter, 2 - polarizing films, 3 - liquid crystal cells, 4 - protective plate.

5.1. Selection of TN Effect for LCS

Selection of the TN electro-optical effect (with the twist angle of $\xi = \pi/2$) for a design of two liquid crystal cells forming LCS [1 - 3] was forced by the requirement of obtaining the protection degree N where $9 < N < 13$ with simultaneous, possibly circularly symmetric, angular characteristics of the Contrast Ratio CR within whole view field. Moreover, this was also implied by the rigorous Power Consumption coefficient PC (PC < 8 $\mu W/cm^2$). The controlling signals with amplitudes U were limited to max. 4.5 V and frequency f $= 50$ Hz). Switching time to the blocking on-state of the filter should be as short as $t_S \sim t_{on} < 0.2$ ms at elevated pulse control voltage U (so-called "overvoltage") in the first control pulse not higher than 80 V.

5.1.1. PC Coefficient of Power Consumption

To determine the power consumption while LCS control and driving one can assume that a replacement system is equivalent to a lossy capacitor which is represented as a parallel-connected system of an ideal capacitor with the high-value resistance R. Here R stands for the resistance of outputs and is responsible for the energy losses for Joule's heat. In liquid crystal cells, there are inevitable losses connected with the dielectric absorption. Power consumption coefficient PC per unit cell area determines the total power P needed for the director of the liquid crystalline slab:

$$PC = \frac{W}{S} = \frac{U \cdot I}{S} = \frac{I^2 \cdot R}{S},$$ (4)

where: I is the electric current intensity passing through the cell of a liquid crystal filter, and R is the impedance of the cell.

The impedance R of a liquid crystal cell filter with the surface area S, the specific resistance ρ and the thickness d of the MNLC mixture slab driven by AC voltage $U(\omega)$ (with the angular frequency $\omega = 2\pi f$ where f is a frequency of controlled voltage) can be written as:

$$R = R_1 + jR_2,$$ (5)

where

$$R_1 = \rho \frac{d}{S},$$ (6)

R_1 is the resistance and j is the imaginary unit ($j^2 = -1$). Reactance R_2 depends on the capacity C:

$$R_2 = \frac{1}{\omega C}.$$ (7)

At on-state the controlling electric field E experiences $\varepsilon_\parallel$ - the parallel component of the real part of the permittivity tensor ε of MNLC mixture:

$$\varepsilon_\parallel = \varepsilon_\perp + \Delta\varepsilon,$$ (8)

where $\varepsilon_\perp$ is the perpendicular component of the real part of tensor ε, and $\Delta\varepsilon$ is the permittivity anisotropy of MNLC mixture so, the capacity C of an LCC is:

$$C = \varepsilon_o(\varepsilon_\perp + \Delta\varepsilon)\frac{S}{d}.$$ (9)

The specific resistance ρ of a well purified liquid crystal (without ionic impurities) is usually not lower than $1 \cdot 10^{12}$ Ωcm.

Let us consider a LCC with electrode surface of $S = 5 \times 10 = 50$ cm^2 and cell gap $d = 6$ μm filled with MNLC mixture of $\rho = 1 \cdot 10^{12}$ Ωcm, $\varepsilon_\perp = 3.0$ and $\Delta\varepsilon = 3.5$. The resistance of such a transducer is $R_1 = 12.0$ MΩ and is much higher than the reactance of the module $R_2 = 0.5$ MΩ. The above causes that power consumption PC of the LCS filter is determined mainly by the reactive current I_2 (displacement current) flowing through an actuated transducer. The current can be written as:

$$I_2 = U\omega C = U\omega\varepsilon_o(\varepsilon_\perp + \Delta\varepsilon)\frac{S}{d}.$$
(10)

This current I_2, at the driving voltage amplitude U not higher than 4.5 V at the frequency of $f = 50$ Hz, should not exceed defined limits and should not be higher than 60 μA).

If one design an MNLC mixture with material parameters like $\rho = 1 \cdot 10^{12}$ Ωcm, $\varepsilon_\perp = 3.0$ and $\Delta\varepsilon = 3.5$ and builds LCC with S $= 5 \times 10 = 50$ cm^2 and $d = 6$ μm then the PC coefficient for the LCS of consisting even of two LCC connected in parallel would not be higher than PC < 8 μ W/cm^2. Such a filter would meet requirements provided for a PC coefficient for that device. At this point, it should be noted that the application of such a filter can allow for the power supply of the welding helmet even by a conventional DC battery with small capacity. (*i.e.*, R6 battery with the capacity of 0.5 Ah can actuate such a helmet with $N = 13$ for 3000 operating hours).

5.1.2. Contrast Ratio CR of the LCS

Contrast ratio CR of the LCS transducer for off-state and on-state is defined as:

$$CR = \frac{T_{OFF}}{T_{ON}},$$
(11)

where T_{ON} is the transmission of LCC at on-state and T_{OFF} in at off-state.

As one can see, the maximum contrast CR is obtained for maximum transmission T_{OFF} at off-state together with minimum transmission T_{ON} at on-state.

It results from the review, and critical analysis of electro-optical TN and ECB effects [4] that subject to the maximum optical contrast is meeting the requirement of optical matching:

$$\text{for TN: } \frac{2d\Delta n}{\lambda} = \sqrt{4k^2 - 1}, \tag{12}$$

$$\text{for ECB: } \frac{2d\Delta n}{\lambda} = 2k - 1, \tag{13}$$

where $k = 1, 2, 3,...$ is the order of the interference maximum of a given effect.

The welder observes his welding spot through a liquid crystal valve (LCS) not only from a regular position (or at the angle of $\theta = 0$) but very often looks at the welding spot along the direction given by observation angles of (θ, Φ). The definition of observation angles (θ, Φ) is presented in Fig. (2).

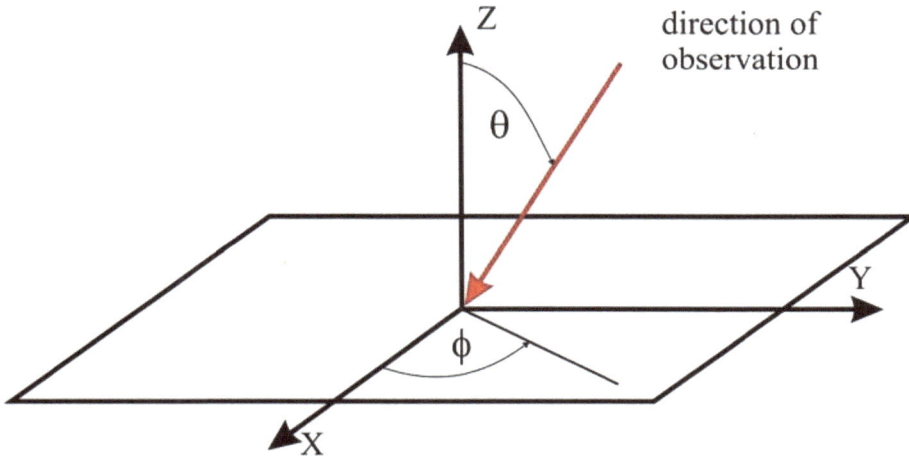

Fig. (2). Direction of observation of the welding spot by a welder defined by the angles θ and Φ.

The liquid crystal cell of the LCS lies in the Oxy plane. The axis Ox coincides with the direction of the director $\boldsymbol{n}$ in the middle of the LC layer. In the case of the ECB effect, the direction of rubbing on both LCS surfaces coincides with the Ox axis. In the case of a TN effect the direction of rubbing forms an angle $\xi_1 = \pi/4$ with the Ox axis on one surface, and an angle $\xi_2 = 3\pi/4$ on the other.

Moreover, sometimes the welding process occurs in a very tight space (*i.e.*, in closed bulkheads of ships or tanks) where many welders work. In such conditions the sight is exposed to harmful radiation from different sources and to scattered light, thus reaching eyes from different directions. Therefore, the angular contrast characteristics $CR(\theta, \Phi)$ of LCS should keep the highest possible contrast within the broadest possible limits.

It results from the analysis of $CR(\theta, \Phi)$ for TN and ECB effects [4] that subject to the maximum contrast CR in a wide range of angles meets the conditions of optical matching (14) and (15) for the smallest possible thickness d. It results from Eqs. (12) and (13) that to successfully protect the welder's sight, the LCS should operate in the first interference maximum ($k = 1$), thus the conditions (12) and (13) can be rewritten as:

$$\text{for TN: } \frac{2d\Delta n}{\lambda} = \sqrt{3}, \tag{14}$$

$$\text{for ECB: } \frac{2d\Delta n}{\lambda} = 1 \tag{15}$$

According to considerations carried out in Section 5.1.1 one can conclude that to meet the technical requirements for LCS power consumption, the liquid crystal cell gap d should not be lower than 6 μm. Taking the above into consideration, as well as knowing that for an optic vision a human eye is the most sensitive to waves of $\lambda = 555$ nm (what we can see in Fig. **3**) the conditions from Eqs. (14) and (15) can be narrowed down to the following considerations for optical anisotropy Δn of nematic mixtures necessary for these applications: $\Delta n = 0.080$ for TN and $\Delta n = 0.046$ for ECB. The development of different MNLC mixtures (described in Chapter 3) presents that the above relatively low optical anisotropies Δn can be obtained in MNLC mixtures consisting of cyclohexylcyanobenzenes mainly and fluorobenzenes strongly diluted.

The dilution process of cyclohexylcyanobenzenes and fluorobenzenes by hydrocarbons causes not only a reduction in the optical anisotropy ($\Delta n < 0.08$) of liquid crystal mixture but also a significant reduction of its permittivity anisotropy $\Delta\varepsilon < 3.5$, which will cause an undesirable increase of filter switching-on time t_{on}.

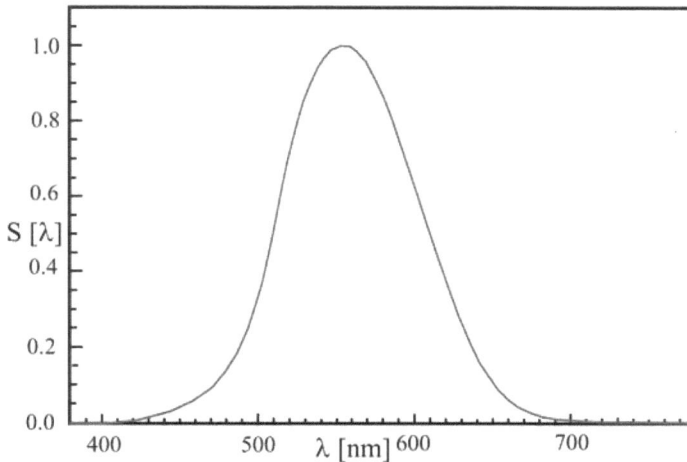

Fig. (3). The diagram of spectral sensitivity for a human eye.

From the considerations in Sections 5.1.1 and 5.1.2, it results that in the LCS of the welding helmet a positive TN mode of a liquid crystal cell of cell gap $d = 6$ μm operating at the first interference maximum should be applied.

5.1.3. TN Effect of the LCS

In Fig. (**4**) the dispersion of the transmittance $T(\lambda)$ of the positive TN mode is presented as a function of wavelength λ for the $d\Delta n = 0.48$ factor calculated based on Eq. (13).

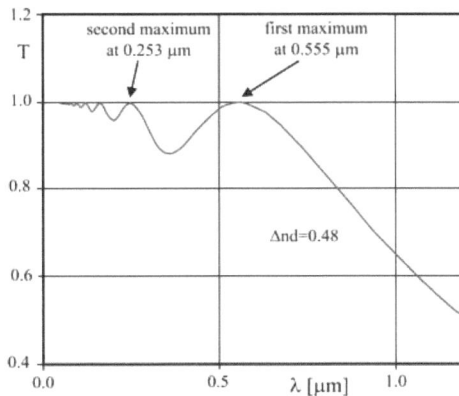

Fig. (4). The dispersion of the transmission $T(\lambda)$ of the LCC operating at positive TN mode calculated based on Eq. (1) in Chapter 4 at $d\Delta n = 0.48$. For MNLC mixture slab of $\Delta n = 0.08$ and $d = 6$ μm, the first interference maximum occurs at $\lambda = 555$ nm - at the wavelength of the maximum of the human eye sensitivity.

In Fig. (**5**) the results [5] of theoretical calculations of optical contrast CR as a function of the angles θ and Φ are presented. The calculations were carried out for $\lambda = 550$ nm of the positive TN mode in a cell with cell gap $d = 5.2$ μm with NLC with $n_e = 1.580$ and $n_o = 1.489$ (or for $d\Delta n = 0.473$ according to the first interference maximum condition). As one can see in Fig. (**5**) the angular characteristics of the contrast CR of one TN cell (by its nature) are highly asymmetrical; thus it was decided to apply two cells in LCS of welding mask. One needs to believe that two TN cells with Ox axes rotated relatively to each other at an angle of $\pi/2$ arranged in series (one after another) in the LCS will make the whole angular characteristics of a CR mask valve near circularly symmetric in relation to the rotation of the angles $\Phi \in [0, 2\pi]$ (Fig. **3**). The calculations reveal that such characteristics should indicate consistent contrast for $\theta \in [0, \pi/4]$ hence the application of two LCC in the LCS should fully meet the requirements for stability of CR within broad limits of the angles θ and Φ.

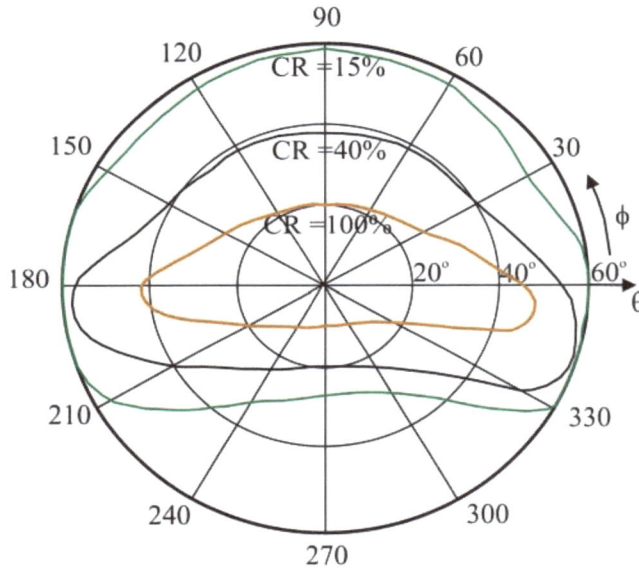

Fig. (5). Results [5] of calculations of Iso-Contrast (CR) in a TN LCC operating at the first interference maximum at $\lambda = 550$ nm. CR(θ, Φ) characteristics were calculated for a cell with thickness $d = 5.2$ μm filled with NLC of $n_e = 1.580$ and $n_o = 1.489$ (or at $d\Delta n = 0.473$).

According to the Tadumi formula [5] (for TN effect) the switching-on time of LCC and hence switching-on time $t_{on} \approx t_{0\text{-}90}$ of the whole LCS can be expressed as:

$$t_{ON} = \frac{\gamma d^2}{(\varepsilon_o \Delta\varepsilon U^2 - \pi^2 K_{TN})},$$

(16)

where K_{TN} is the reduced elastic constant of MNLC mixture for a TN effect with $\xi = \pi/2$ placed in a TN cell with thickness d controlled by alternating voltage with amplitude U.

One can obtain a short switching-on time t_{ON} of LCC using the thinner liquid crystal cell (small d) as well as with using of MNLC mixture which indicates the high value of permittivity anisotropy $\Delta\varepsilon$, the low value of the constant K_{TN} and the low viscosity γ.

Due to the previously discussed restrictions for the cell gap d (which should not be lower than 6 µm) as well as for dielectric anisotropy $\Delta\varepsilon$ (which should not be higher than 3.5) results, that practically the driving voltage value U is the main parameter controlling switching time (t_{on}) of a transducer. The higher is the driving voltage U, the shorter is the time t_{on} (see Eq. (16)). Based on Eq. (16) one can evaluate the switching time t_{on} of LCC with cell gap $d = 6$ µm, driving voltage amplitude U, a liquid crystal mixture with $\gamma = 170$ mPa·s, $\Delta\varepsilon = 3.5$ and $K_{TN} = 16$ pN. At the voltage $U = 4.5$ V the switching time is $t_{on} = 13.0$ ms. An increase of the amplitude U of the first pulse (during control pulses of the signal) to the value of $U = 50$ V reduces the switching time to 0.08 ms. In this case, LCC meet the requirements for switching time t_{on}, which demands voltage value of the first driving pulse not higher than $U = 80$ V and the switching time t_{on} shorter than 0.2 ms.

Implementation, control and selection of liquid crystal TN cells with a relatively high aperture as well as a careful and precise assembly under conditions of a high purity eliminating all inaccuracies allows meeting all requirements for differences under prism power, a heterogeneity of light transmission coefficient and the dispersion.

The previous authors' experiences with the TN effect indicate that LCS based on TN effect, being carefully assembled, also meets the general requirements for a spherical and astigmatic power.

5.2. Selection of MNLC Mixture for LCS

In Section 3 and 5.1 of this chapter the requirements for material parameters of the MNLC mixture for LCS are determined. Such an MNLC mixture should indicate the following values of material parameters at the room temperature:

- specific resistance $\rho > 1 \cdot 10^{12}$ Ωcm,
- rotational viscosity $\gamma < 170$ mPa·s,
- optical anisotropy $\Delta n = 0.08$,

- perpendicular component of permittivity $\varepsilon_\perp < 3.0$,
- dielectric anisotropy $\Delta\varepsilon = 3.5$,
- reduced elastic constant $K_{TN} < 16$ pN.

According to the idea formulated above, the team of chemists from the Institute of Chemistry of the MUT created the working W1115 mixture with the phase sequence: Cr -7.0°C N 60.0°C Iso. Mixture W1115 was thoroughly studied by the team of physicists from the Institute of Applied Physics of the MUT [1, 6].

Detailed results of the examination of the W1115 mixture parameters are presented in Chapter 3 and selected material parameters of this mixture, useful for LCS, are gathered below, in Table **1**.

All material parameters of the W1115 mixture gathered in Table **1** meet the requirements formulated in Section 6.5.1. The temperature range of the nematic phase (from -7.0°C to +60.0°C) is much broader than the temperature range of the nominal work of the welding helmet. Material parameters W1115 are almost constant within operating temperatures (from 0°C to +35°C).

Table 1. Material parameters of the liquid crystal mixture W1115 useful for LCS (assessed at 25°C).

Parameter	Value, Unit
Specific resistance ρ	$1.3 \cdot 10^{12}$ Ωcm
Optical anisotropy Δn ($\lambda = 589$ nm)	0.08
Optical anisotropy Δn ($\lambda = 1064$ nm)	0.07
Ordinary refractive index n_o ($\lambda = 589$ nm)	1.56
Dielectric anisotropy $\Delta\varepsilon$ ($f = 1.5$ Hz)	3.5
The perpendicular component of permittivity tensor $\varepsilon_\perp$	3.0
Splay deformation elastic constant K_{11}, (U_{th})	10.2 pN, (U_{th}=1.8 V)
Reduced elastic constant K_{TN}, (U_{TN})	15.2 pN, (U_{TN}=2.2 V)
Rotational viscosity γ	165 mPa·s

5.3. Structure of Liquid Crystal Cells for LCS

The structure of a liquid crystal cell of the LCS is presented in Fig. (**6**).

To develop LCCs typical processes and laboratory, devices were applied, allowing precise montage of the element of the LCS. Major directions of technological works focused on extending hardware capabilities by modifications allowing to achieve desired results. Processes and materials having a decisive influence on cell gap and homogeneity of a liquid crystal slab underwent a modification. Sodium glass plates of the thickness of 1.1 mm and flatness higher then 0.1 μm / 25mm were used as filter substrates. This is typical float glass applied at the production of liquid crystal displays. Glass was equipped with ITO electrodes of a specific resistivity c.a. 100 Ω/□ patterned while a wet etching process. Electrodes were covered with SiO$_2$ barrier layer.

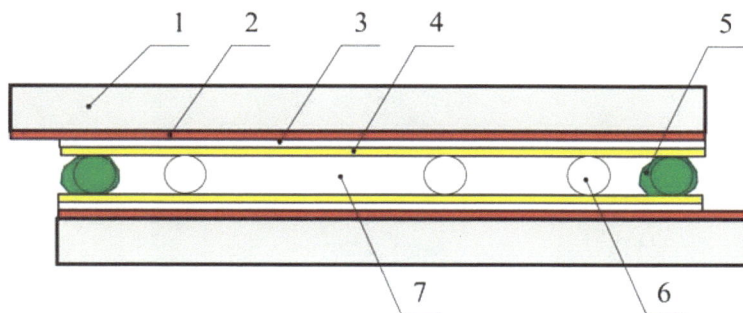

Fig. (6). Structure of liquid crystal cell: 1 - glass housing plates, 2 - electrode layers In$_2$O$_3$: Sn, 3 - insulating layers SiO$_2$, 4 - polyamide orienting layers, 5-spacing-encapsulating layer, 6- glass, 7-liquid crystal layer.

To increase the cell gap and cell gap uniformity control, a low viscosity photocurable resin UVS 91 (by Norland Adhesives) was used. For sealing the plates a gradient-less press, providing homogeneous conditions of the cell assembly, was utilized. Cell gap uniformity was preserved with using of spacers in the form of microspheres distributed over a glass surface. It was evaluated through observation of interference fringes in a monochromatic illumination of sodium vapor lamp. Observation of interference fringes confirmed the flatness of obtained cells. Satisfactory result of the inspection (not more than two fringes observed with symmetric distribution, reproducing the shape of the outer edge) was pre-requisite to take up the hardening process by UV lamp irradiation. After the resin hardening, final flatness control was conducted.

Positive results at previously used criteria were the condition of admission to the next stages of the technological process. The final cell gap d of LCC was determined by interference measurements conducted using a JASCO V670 spectrometer. The final quality classification was made after filling the cells with the W1115 mixture. Cells filling with liquid crystal material, lead by capillary action, improved the homogeneity of the cell gap. The homogeneity of a liquid crystal material alignment is a derivative of a liquid crystal material type and liquid crystal interaction with a specific orienting material as well as the method of treating the aligning surface. The quantitative appraisal of homogeneity as well as of the quality of a liquid crystal material alignment was made based on an observation of the MNLC texture under the polarizing microscope.

5.4. Permanent Filters

To meet requirements of sight and equipment protection, the shape of the dispersion of the transmission $T(\lambda)$ of the optical elements which constitute a protective device should be tailored to protect sight within VIS range but also completely suppress radiation in the spectrum areas outside VIS range; in the ultraviolet (from 100 nm to 420 nm) and the infrared part of the spectrum (700 nm to 12000 nm). Such dispersion characteristics can be obtained through:

- vacuum vapor-deposited stacks of dielectric layers (Dielectric Interference Filters DIF),
- vacuum vapor-deposited stacks of dielectric-metallic nature (Dielectric-Metallic Interference Filters DMIF),
- use of dyed plates (with suitable dyes) of mineral or polymer glass (Dyed Filters DF).

The use of the absorption phenomenon in DF has two advantages concerning interference filters. Firstly, it shows no adverse reduction in transmission T with increasing incident angle θ. Moreover, with the increase of incident angle θ, the distance in which radiation travels through plates gets longer, which increases the attenuation. Secondly, interference filters, as a general rule, reflect radiation which has to be blocked what is not always desirable. In Fig. (**7**) the transmission spectrum $T(\lambda)$ for a dyed filter (DF) is presented. This kind of filter developed by MASKPOL LTD is applied in gas-proof masks of MP6 type.

The major defect of DF filters at considered application is their significant thickness, usually much higher than 5 mm, thus a significant weight.

The example of transmission spectrum $T(\lambda)$ of the dielectric interference filter within the VIS range is presented in Fig. (**8**).

In order to obtain transmission spectrum $T(\lambda)$ presented in Fig. (**8**) one needs to vaporize the filter substrate with dozens of dielectric layers (*i.e.*, alternately TiO_2 and SiO_2) with specific thicknesses what makes this technology unhelpful to welding masks mass production.

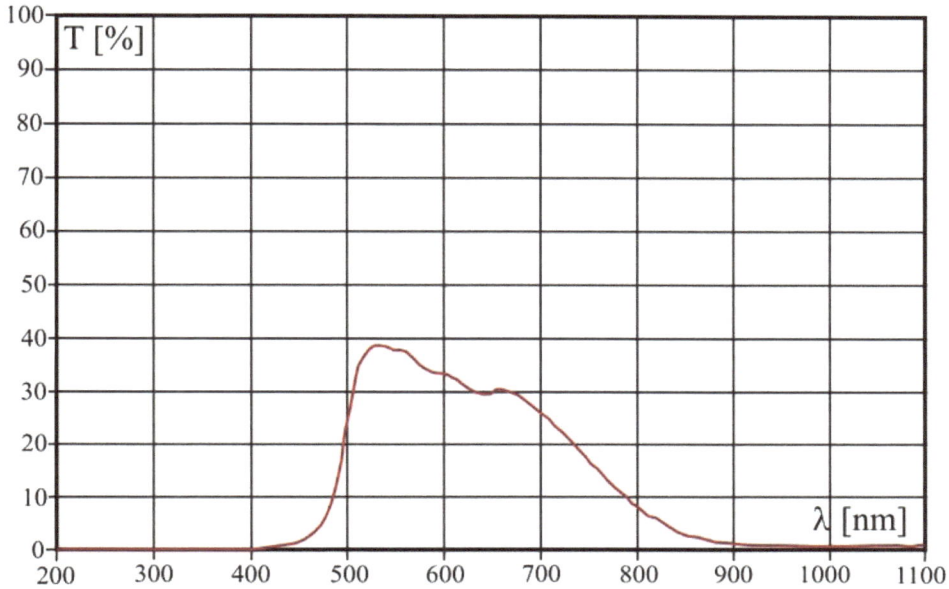

Fig. (7). Dyed Transmission Filter manufactured by MASKPOL LTD for gas-proof MP6 masks.

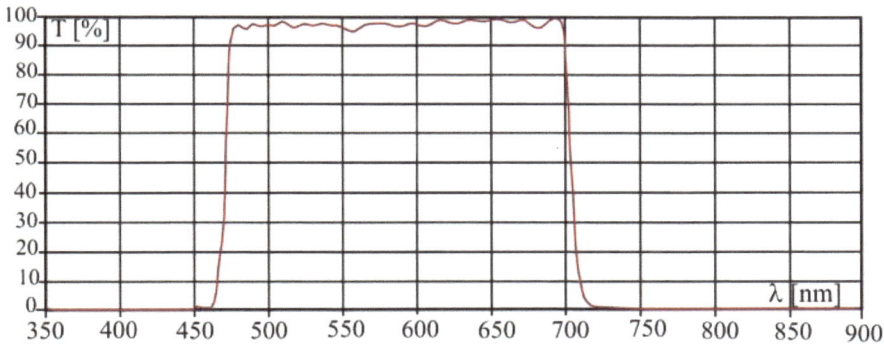

Fig. (8). Transmission of Dielectric Interference Filter - DIF developed at the MUT (consisting of dozens of TiO_2-SiO_2 layers).

In Fig. (9) the transmittance spectrum $T(\lambda)$ of dielectric-metallic interference filter (DMIF) at the VIS range is presented. From the spectral transmittance spectrum $T(\lambda)$ presented in Fig. (9), one can evaluate the total transmission ratio $\tau_F \sim 0.31$ for DMIF by using Eq. (6). This kind of filter developed at the MUT was applied in an automatic welding helmet (PIAP PS automatic) meeting requirements of Central Institute for Labor Protection (CIOP in Polish) for protection degree ranges from 9 to 13 N.

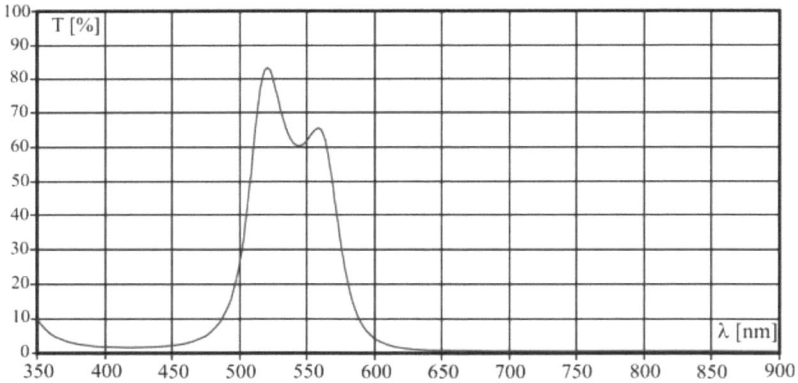

Fig. (9). Transmission of Dielectric-Metallic Interference Filter DMIF developed and applied at the Optoelectronics Institute of the MUT.

5.5. LCS of a Welding Mask

After cells manufacturing and filling them with the liquid crystal, W1115 mixture LCC was taped with polarizer NPF-Q-12-35 (by NITTO) and stacked. Subsequently, the stack was equipped with the interference (dielectric-metallic) permanent filter DMIF and a protection plate (see Fig. **1**).

In Fig. (**10**) the transmittance spectrum $T(\lambda)$ of polarized films NPF-Q-12-35 (by NITTO) applied in LCS is presented. From the transmittance spectrum $T(\lambda)$ (see Fig. **10**) one can evaluate the total transmission ratio by using Eq. (2). It was evaluated as $\tau_p \sim 0.410$ for a single film and as $\tau_{POFF} \sim 0.300$ for two polarized films when the polarization axes of polarizers are parallel and as $\tau_{PON} \sim 0.004$ when the axes are perpendicular.

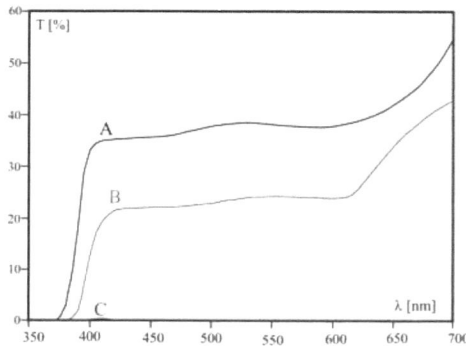

Fig. (10). Transmittance spectrum $T(\lambda)$ of polarized films NPF-Q-12-35 (by NITTO) applied in LCS. A - single polarized film, B - two films with parallel polarized axes, C - two films with perpendicular polarization axes.

In Fig. (**11**) the transmittance spectrum $T(\lambda)$ of a single cell of LCS, placed between two crossed polarizers, at off-state (when $U = 0$ V) is presented. In this figure, one can see that the first interference maximum of a TN in this cell (6.1LCCF) occurs at the wavelength $\lambda = 555$ nm (compare with Fig. **3**).

The above provides a good optical fit of the W1115 mixture for work at the first interference maximum ($k = 1$, $d\Delta n = 0.48$) of a TN effect for the maximum at $\lambda = 555$ nm.

From the transmittance spectrum $T(\lambda)$ placed in Fig. (**11**), by using Eq. (6), one can evaluate the total transmission ratio $\tau_{KOFF} \sim 0.75$ for a single liquid crystal cell of LCS.

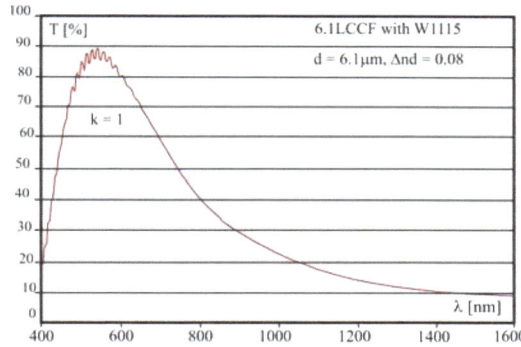

Fig. (11). The spectrum of the transmission $T(\lambda)$ of 6.1LCCF cell (of a cell gap $d = 6.1$ μm and with transparent ITO electrodes of the specific resistance of 70Ω/□) filled with W1115 ($\Delta n = 0.08$ at $\lambda = 589$ nm) at 25°C placed between crossed polarizers. The polarization plane of incident light was perpendicular to the director n in a TN layer at the entrance to 6.1LCCF.

If one could build an LCS consisting of a single cell (with $\tau_{KOFF} \sim 0.75$) placed between two crossed polarizers ($\tau_{POFF} \sim 0.30$) and an interference filter DMIF (with $\tau_F \sim 0.31$), the total transmission ratio of such a filter at the off-state would be $\tau_{OFF} = \tau_{KOFF} \cdot \tau_{POFF} \cdot \tau_F \sim 0.07$. According to Eq. (1) this filter in an off-state will have $N = 2.7$. In an on-state (for which we can assume that the transmission ratio of LCC equals $\tau_{KON} \sim 1$), $\tau_{ON} = \tau_{KON} \cdot \tau_{PON} \cdot \tau_F \sim 0.00124$, what according to Eq. (1) provides a protection degree of $N = 7.8$. In such a situation, it was decided to build an LCS consisting of two liquid crystal cells placed between three crossed polarizers as it is presented in Fig. (**1**). In this target filter, we should obtain:

at off-state $\tau_{OFF} = \tau_{KOFF} \cdot \tau_{POFF} \cdot \tau_{KOFF} \tau_P \cdot \tau_F \sim 0.021$ what makes $N = 4.2$, and

at on-state $\tau_{ON} = \tau_{KON} \cdot \tau_{PON} \cdot \tau_{KON} \cdot \tau_{PON} \cdot \tau_F \sim 0.000005$ what makes $N = 13.4$.

Thus, one can see that the LCS consisting of two liquid crystal cells (with cell gap

$d = 6$ μm filled with the W1115 mixture of $\Delta n = 0.08$ at $\lambda = 589$ nm) placed between three crossed polarizers NPF-Q-12-35 and a band-pass filter DMIF fully meets the requirements to be applied in professional welding helmets with protection degree up to 13 N.

In Fig. (**12**) an angular contrast characteristics $CR(\theta, \Phi)$ of the LCS consisting of two liquid crystal 6.2LCCF cells ($d = 6.2$ μm) placed between three crossed polarizers at the directors **n** in the middle of 6.2LCCF is presented. Cells are rotated to each other about the angle of $\pi/2$. One can see that this total angular characteristics $CR(\theta, \Phi)$ of the LCS valve of the welding mask shows almost circular symmetry concerning a rotation angle $\Phi \in [0, 2\pi]$. Moreover, these characteristics $CR(\theta, \Phi)$ maintains a constant contrast (the same as for $\theta = 0$) within the range required by appropriate standards, which is $\theta \in [0, \pi/4]$.

In Fig. (**13**) the process of measurement of a switching-on t_{on} and off t_{off} times of the welding mask LCS's valve is illustrated. If the filter is controlled by a sequence of rectangular pulses wave (with a length of 0.5 s) with an amplitude of $U = 4.5$V and a frequency of $f = 50$ Hz (see Fig. **13a**), switching-on time t_{on} equals 13 ms, and switching-off time is $t_{off} = 35$ ms (see Figs. **13b** and **13c**). This switching-on time is the same as the time evaluated in Section 5.1.3 for the liquid crystal mixture searched for this filter. The dependence of switching times as a function of driving voltage is presented in Table **2**.

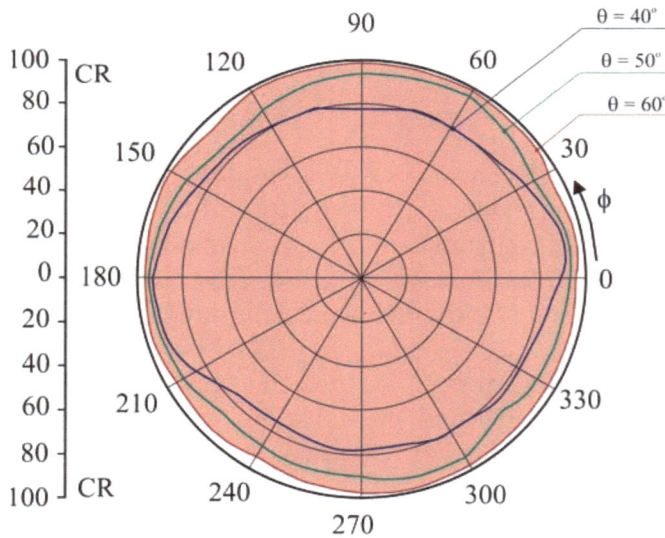

Fig. (12). Experimental results of Iso-Contrast (CR) in a liquid crystal filter consisting of two liquid crystal cells of 6.2LCCF (d=6.2 μm) placed between three crossed polarizers (see Fig. 1). Directions of the director **n** in the middle of 6.2LCCF cells are rotated to each other about an angle of $\pi/2$).

Table 2. Switching-on times t_{ON} of the LCS with W1115 mixture at 25°C.

U [V]	4.5	10.0	20.0	40.0	50.0	60.0	80.0	100.0	120.0
t_{ON} [ms]	13.00	2.60	0.67	0.19	0.11	0.09	0.07	0.05	0.04

In order to decrease switching-on time, switching time into blocking time t_S (at switching pulse not higher than $U = 80$ V) should not be higher than $t_S = 0.2$ ms while reducing power consumption in control, the short first pulse was applied of an amplitude $U = 50$ V (see Fig. **14**) during control what allowed to meet both requirements.

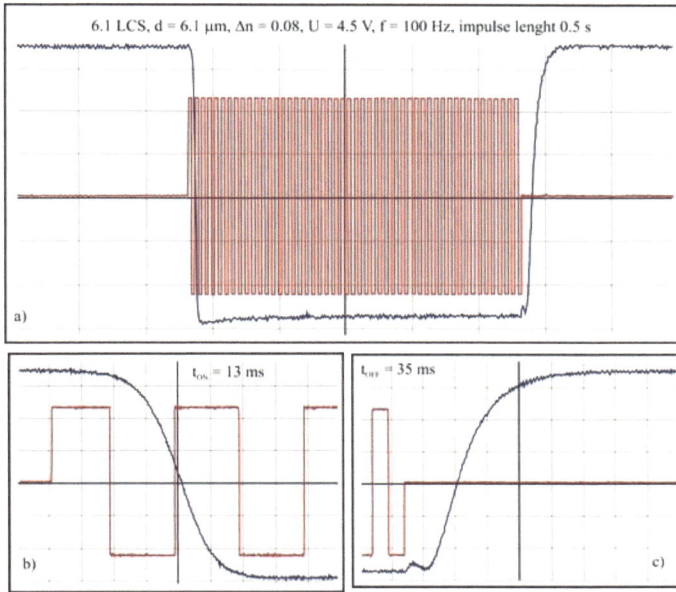

Fig. (13). Oscillograms of switching-on t_{ON} and off t_{OFF} times of the welding mask' valve of LCS. Description in the text.

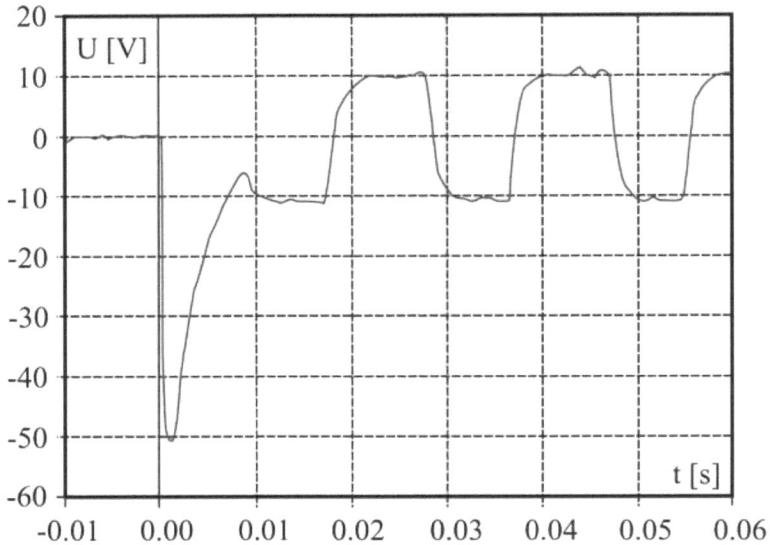

Fig. (14). Electric signal fed to the electrodes of the LCS of the "PIAP-PS automatic" welding helmet after the flash of light emerging at a transition time of 20 ns.

CONCLUSION

The study of helmets with LCS carried out in Central Institute for Labor Protection (CIOP in Polish) confirmed the above results of MUT's tests. In 1997, CIOP issued the appropriate safety certificate (No. 244/98 CIOP and 245/98 CIOP) allowing for the usage of Automatic Welding Helmet (PIAP-PS automatic) in Poland.

Table 3. Technical parameters of the automatic welding helmet (PIAP-PS automatic) with LCS.

Parameter	Value, Unit
Dimensions of the active windows	100 mm x 43 mm
Range of protection adjustment	9N-13N
Blackout degree in "OFF" state	4N
Blackout switching time	< 0.2 ms
Range of operating temperatures	from -5°C to +60°C
Power source	Solar cell
Weight (without headband)	370 g

This helmet was manufactured in series in 1998-2005 (100 pieces/month) by Industrial Research Institute for Automation and Measurement in Warsaw,

Poland. Specification of parameters of the automatic welding helmet PIAP-PS automatic are gathered in Table **3**.

The helmet was award-winner; it was rewarded with Gold Medal at "BRUSSELS EUREKA'94", Gold Medal at "International fair in Poznan'94", Grand Prix SAWO'93 at "International Fair of Work Protection, Fire-Fighting and Risk Equipment", Award NASZ DOM (Our Home)'92 and was nominated for the award TERAZ POLSKA (Poland Now) at the 2nd edition.

CONSENT FOR PUBLICATION

Not applicable.

CONFLICT OF INTEREST

The author(s) confirms that there is no conflict of interest.

ACKNOWLEDGEMENTS

Declared none.

REFERENCES

[1]   E. Nowinowski-Kruszelnicki, "Optical element for automatically darkening welding filters", *Proc. SPIE,* vol. 3318, pp. 519-522, 1998.
        [http://dx.doi.org/10.1117/12.300038]

[2]   M. Schadt, and W. Helfrich, "Voltage-dependent optical activity of a twisted nematic liquid crystal", *Appl. Phys. Lett.,* vol. 18, no. 4, pp. 127-128, 1971.
        [http://dx.doi.org/10.1063/1.1653593]

[3]   C.H. Gooch, and H.A. Tarry, "The optical properties of twisted nematic liquid crystal structures with twist angle < 90 degrees", *J. Phys. D Appl. Phys.,* vol. 8, no. 13, pp. 1575-1584, 1975.
        [http://dx.doi.org/10.1088/0022-3727/8/13/020]

[4]   P. Yeh, C. Gu,, "*Optics of Liquid Crystal Displays*", A Wiley Interscience Publication, John Viley & Sons, Inc., New York/Chichester/Weinheim/Brisbane/Singapore/Toronto, (1999), ISBN OA71-18201-X.

[5]   K. Tarumi, U. Frinkenzeller, and B. Schuler, "Dynamic behaviour of twisted nematic liquid crystals", *Jpn. J. Appl. Phys.,* vol. 31, pp. 2829-2836, 1992.
        [http://dx.doi.org/10.1143/JJAP.31.2829]

[6]   E. Nowinowski-Kruszelnicki, J. Kędzierski, Z. Raszewski, L. Jaroszewicz, R. Dąbrowski, W. Piecek, P. Perkowski, K. Olifierczuk, K. Garbat, M. Sutkowski, E. Miszczyk, K. Ogrodnik, P. Morawiak, M. Laska, and R. Mazur, "High birefringence liquid crystal mixtures for lc electro-optical devices", *Opt. Appl.,* vol. 42, no. 1, pp. 167-180, 2012.

LIST OF ABBREVIATIONS

CONTRACTUAL DESIGNATIONS AND SYMBOLS USED IN THE PAPER

E — intensity vector of external electric field,

E_C — threshold value (critical) of intensity of external electric field,

H — intensity vector of external magnetic field,

H_C — threshold value (critical) of intensity of external magnetic field,

B — induction vector of external magnetic field,

B_C — threshold value (critical) of induction of external magnetic field,

U — voltage applied to electrodes,

U_C — threshold value (critical) of voltage applied to electrodes,

f — frequency,

ω — wheel frequency, angular frequency,

δU_C — absolute evaluation error of threshold value of a voltage applied to electrodes,

NLC — Nematic Liquid Crystal,

T — Transmittance (energy transmission coefficient), sometimes temperature,

τ — Transmission coefficient,

t_{iji}) — amplitude ratio of a transmitted wave (to a medium with n_j) to the incident wave (from a medium with n_i),

R — Reflectance (energy reflectance),

ρ — Reflectivity coefficient,

r_{ij} — amplitude ratio of reflected wave (to a medium with n_j) to the incident wave (from a medium with n_i),

A — Absorption (energy absorption coefficient),

α — Absorption coefficient,

I — isotropic wave sometimes wave intensity after passing through the medium,

I_0 — intensity of incident wave at the medium,

a — wave amplitude,

φ — wave phase,

$\Delta\varphi$ — phase difference between the wave components,

F — coefficient of finesse "focus",

N — Nematic phase,

TN — electro-optical Twisted Nematic effect,

ξ — twisted layer angle in a TN effect,

ECB — Electric Controlled Birefringence electro-optical effect

$\textbf{VAN}$ electro-optical effect (Vertical Aligned Nematic),

Φ aperture,

$\textbf{SmA}$ smectic phase A,

$\textbf{MNL}$ Nematic Liquid Crystal Mixture,

K_{11} Frank elastic constant describing S deformation (splay),

δK_{11} absolute error of elastic constant K_{11} determination,

K_{22} Frank elastic constant describing T deformation (twist),

δK_{22} absolute error of elastic constant K_{22} determination,

K_{33} Frank elastic constant describing B deformation (bend),

δK_{33} absolute error of elastic constant K_{33} determination,

K_{TN} reduced elastic constant for a TN effect,

$\textbf{ITO}$ transparent conductive layer with ITO (Indium Tin Oxide),

$\textbf{PITO}$ porous, transparent conductive layer with ITO,

$\textbf{QP}$ plane parallel quartz plate (Quartz Plates),

$\textbf{AR}$ antireflective layer,

$\textbf{BF}$ blocking layer,

$\textbf{PL}$ alignment layer,

ε_0 vacuum permittivity,

$\varepsilon_{\parallel}$ parallel component of permittivity tensor NLC,

$\varepsilon_{\perp}$ perpendicular component of permittivity tensor NLC,

$\Delta\varepsilon = \varepsilon_{\parallel}\text{-}\varepsilon_{\perp}$ permittivity anisotropy of NLC,

μ_0 vacuum magnetic permeability,

$\Delta\chi$ anisotropy of magnetic susceptibility of NLC,

λ light wavelength,

$\textbf{VIS}$ range of visible wavelengths (VISible),

$\textbf{NIR}$ range of near-infrared (Near Infra Red),

$\textbf{n}_e$ extraordinary refractive index of the NLC,

$\textbf{n}_o$ ordinary refractive index of the NLC,

$\Delta\textbf{n} = \textbf{n}_e - \textbf{n}_o$ optical anisotropy NLC,

$\textbf{n}$ NLC layer director,

$\textbf{n}(r)$ vector field director in a NLC layer,

$\textbf{W}$ anchoring energy of the director NLC on boundary wall,

$\textbf{FoM}$ quality factor (Figure of Merit),

$\textbf{NLC}$ NLC with $\Delta\varepsilon < 0$ (Negative Liquid Crystal),

$\textbf{PLC}$ NLC with $\Delta\varepsilon > 0$ (Positive Liquid Crystal),

LCD	liquid crystal (indicator) filter (Liquid Crystal Display),
LCS	Liquid Crystal Shutter,
LCDTV	liquid crystal TV set,
3DLCDTV	liquid crystal TV set with stereoscopic image (3D),
5CB	4-4'-n-pentyl-cyanodiphenyl,
6CHBT	4-trans-4'-n-hexyl-cyclohexyl-isothiocyanatobenzene,
MBBA	4-(trans-4'−n-hexylcyclohexyl)-isothiocyanatobenzene,
DE	NLC being a mixture of two Demus esters (Demus Esters),
TBA	Tilt Bias Angle,
d	measurement cell thickness,
δd	absolute error of measurement cell thickness designation,
IPS	In Plane Switching,
b	width of IPS electrode (strip),
l	width of gap between IPC electrodes' strips
δl	absolute error of designation of IPS electrodes' strips gap width,
c=l+b	raster of an IPS electrode.

SUBJECT INDEX

A

Abbe refractometer 36, 37, 39, 49, 55, 57, 58, 61, 63
Absorption 77, 80, 83, 84, 85, 86, 93, 115, 129
   coefficient 80, 115
   dielectric 119
   phenomenon 129
Adaptive optics 2
Air pollution(s) 1, 2, 97, 106, 107, 112
   detection 1
   diagnostics 112
   diagnostics unit 106, 107
   level 97
Alignment 2, 4, 11, 19, 33, 89, 91
   homogeneous (HG) 9, 13, 14, 33, 34, 99,
   homeotropic 9, 14, 20, 33, 34,
   hybrid 4, 9,
   layers 14, 33, 34, 83, 85, 89, 108
   techniques 2
Angular frequency 76, 120
   antireflective layer (AR) 79, 87, 88, 89, 90, 91, 92,
Automatic liquid crystal shutters 115

B

Beam 14, 38, 68, 69, 85, 88, 70, 77
   incident 70
   laser 38, 68, 69, 85, 88
   polarized 14
Beam's energy density 68
Bend
   deformation 3, 9, 19, 20, 41,
   elastic contact 3, 9, 25, 40, 72,
Blackout degree 135
Blocking
   film (BF) 87, 88, 89, 90, 91, 92
   layers BF 87, 89

C

Characteristics 31, 32, 81, 115, 118, 119, 125
   angular 119, 125
   electro-optical 32
   material 31
   non-linear 115
   spectral 81, 115, 118
Characteristics CR 123, 133
   angular contrast 123, 133
   total angular 133
Circular symmetry 133
Conditions
   boundary 8
   normal illumination 114, 115
   regular illumination 116
Conductive layers 33, 79
   transparent 79
Conjugate amplitude AR 78
Constant
   elastic 3, 9, 10, 13, 14, 23, 27, 39, 41-47, 65, 66, 72, 74, 75, 103, 108, 126, 127
   reduced 44, 46, 47, 72, 126, 127

D

Dielectric 3, 27, 32, 35, 64, 68, 70, 77, 78, 79, 81, 88, 89, 96, 114, 119, 129
   absorption 119
   anisotropy 3, 10, 23, 27, 68, 70, 75, 93, 96, 108, 114, 126, 127,
   filters 118, 129
   interference Filters DIF 129
   layer 79, 89, 129
Dielectric-metallic 114, 129, 130, 131
   band-pass filter 133
   interference filter (DMIF) 114, 129, 130, 131, 132
   permanent filter 131
Dye 129, 130
   filter (DF) 129
   transmission filter 130
Dynamics, switching 73

E

Elastic constants 3, 9, 10, 13, 14, 23, 27, 39, 41-47, 65, 66, 72, 74, 75, 103, 108, 126, 127
Electrically controlled birefringence effect (ECB) 46, 47, 65, 108, 121 - 123,
Electrically tunable liquid crystal filter 96
Electric field 8, 9, 11, 12, 13, 16, 19, 20, 23, 24, 25, 99, 106, 108
  action 99
Electrodes 4,12, 13, 14, 15, 16, 19, 21, 23, 24, 25, 26, 33, 34, 40, 41, 42, 43, 49, 52, 88, 97
  direction 13
  inter-digital 4, 12, 14
  interdigitating 13, 15, 24
  planar 13
  solid 21, 23, 25, 26
  transparent 25, 88, 97
Electro-optical effects 1, 8, 94
Electro-optic welding lens assembly 115
Ellipsoid 15, 16, 97
Energy 2, 80
  conservation 80
  consumption 2
Evaporated gold-on-chrome layers 89
Evaporation 26
Eye protection 66, 115
  devices 66, 115
  transducers 115

F

Films 118, 119, 131
  polarized 131
  polarizing 119
  polarizing polymer 118
Filter 96, 112, 115, 116, 117, 118, 119, 121, 123, 128, 129, 130, 132, 133
  automatic visible light 112
  dielectric interference 129, 130
  optical 116
  permanent protecting 117
  protection 117
  substrates 128, 129
  target 132
  tunable 96
Flatness 85, 128

Float glass 25, 33, 84, 108
  substrate 25
Freedericksz's transitions 8, 9, 28, 41

G

Geometries 10, 11, 19, 25, 26
  of fields 26
  of NLC 10, 11
Glass 33, 70, 91, 128
  commercial flat 91
  high-quality float 33
  plates 33, 84
  polymer 129
  spacers 34, 89, 109
  substrates 70
  surface 33, 128

H

Homogeneity 12, 13, 14, 128, 129
  conditions 13, 14
  of deforming 12, 13
Human eye sensitivity 124
Hybrid 8, 24, 25, 27, 28, 58
  cell 27, 28, 58
  In-Plane-Switched (HIPS) 8, 24, 25

I

Illumination 114, 115, 118
  intensive 114, 115
  intensity 118
Image 2, 115
  analysis 2
  intensifiers 115
  transducer 115
Indium tin oxide (ITO) 33, 70, 79, 84, 88, 91, 97
  electrodes 70, 34, 83, 84, 103, 104, 108, 128, 132
  layer 15, 25, 83, 84, 86
  transparent 70
Induction 23, 108
  of homogeneous electric fields 23
In-Plane Switching (IPS) 4, 8, 12, 14
  type 12
Interference 37, 48, 49, 51, 52, 55, 57, 58, 60, 76, 81, 129, 131

www.ingramcontent.com/pod-product-compliance
Lightning Source LLC
Chambersburg PA
CBHW041711210326
41598CB00007B/609